Calculus Explorations with *Geometry Expressions*™

Irina Lyublinskaya

CUNY College of Staten Island, New York, NY, USA

Valeriy Ryzhik

Lycee Physical -Technical High School, St. Petersburg, Russia

Ron Armontrout

The Hotchkiss School, Lakeville, CT, USA

Saltire Software
P.O. Box 230755
Tigard, OR 97281-0755
http://www.geometryexpressions.com/
http://www.saltire.com/
support@saltire.com

Table of Contents

Introduction

Teachers know the difficulties in motivating many students to develop the habits of mind and critical thinking skills necessary to thoroughly understand the concepts of calculus. The purpose of this book is to use *Geometry Expressions* software in order to facilitate and enhance the calculus syllabus by allowing students to ground calculus concepts in a geometric way.

Major calculus concepts, such as derivative and integral of a function, have a clear geometric meaning. This encourages students to visualize the concepts and make connections between its geometric and algebraic representations. For example, a function can be represented geometrically by its graph; the derivative of the function is visually represented by the slope of the tangent line; the definite integral of the function is an area under the graph of the function. These geometric representations serve as a basis for a conceptual introduction of the concepts of derivative and integral. The formal definitions of the calculus concepts then lead to clarification of these geometric ideas.

For example Exploration 2.1, "Exploring Tangent Lines", introduces the conceptual idea of the derivative. After the derivative is formally defined, it is possible to interpret the slope of the tangent line in terms of the derivative. Similarly in Exploration 4.5, "The Trapezoid Method", an area of a curvilinear trapezoid is used to develop the concept of a definite integral. After an introduction of the Fundamental Theorem of Calculus, the area can be defined using a definite integral. Both of these examples are based on the concept of the differential of a function which has a clear geometric meaning, as shown in Figure 1.

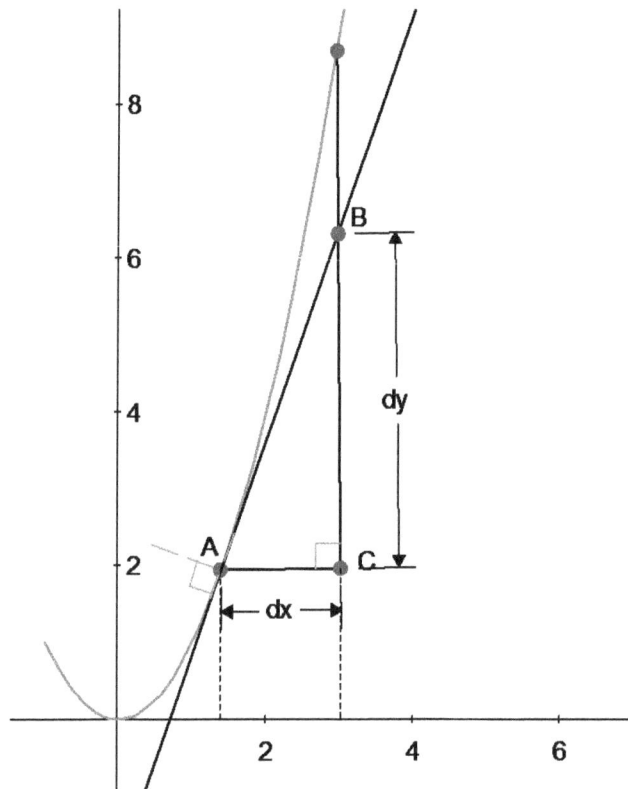

Figure 1. The derivative of a function at point A is a ratio of the sides of right triangle ABC

Not only do the main concepts of calculus have geometric meaning, but also its most important theorems can be interpreted geometrically.

1. The family of antiderivatives of a given function can be found by the translation of a particular antiderivative along the *y*-axis.

2. Derivatives of basic operations can be derived using the idea of areas as shown in Figure 2 for an example of the derivative of a product

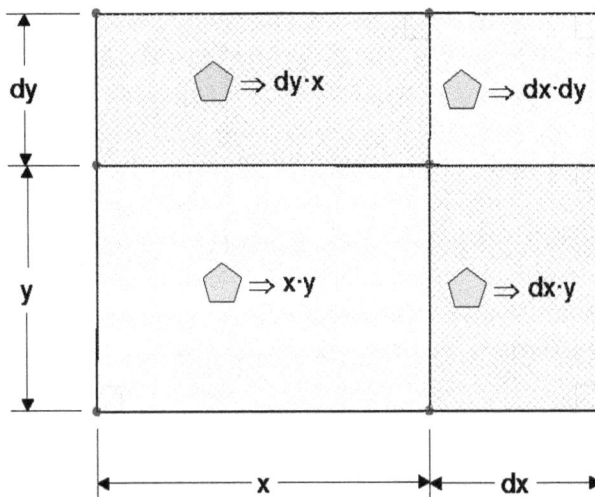

Figure 2. Derivative of a product using area.

3. Finding extremum of a function is equivalent to finding a point on the graph of a function, where the tangent line is horizontal (if that point exists).

4. The Mean Value Theorem determines a point on a graph of a function, defined on an interval, such that the tangent line at this point is parallel to the secant line through the endpoints of the interval.

5. Approximation methods, such as Newton's method for finding the zeros of functions or numerical methods for finding areas are based on clear geometric ideas (see Figure 3).

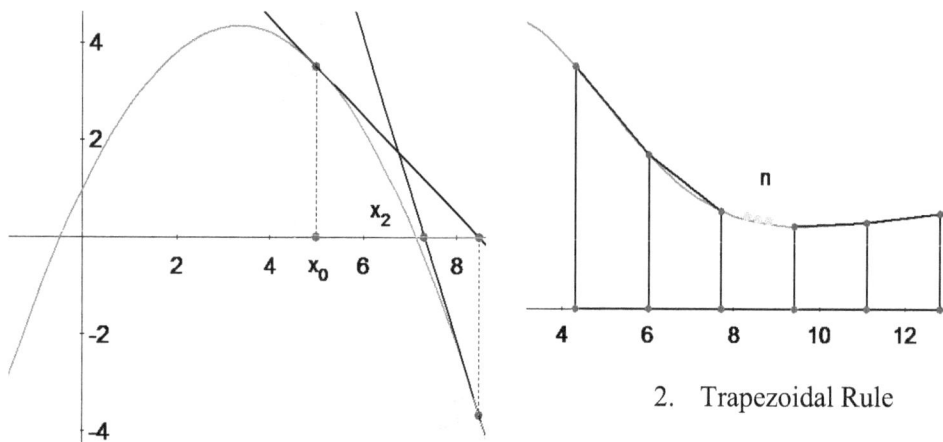

1. Newton's Method

2. Trapezoidal Rule

Figure 3. Illustrations for approximation methods in calculus

6. The concavity of a function is defined geometrically based on the fact that all tangent lines are above or below the graph of a function.

7. The limit and asymptote of a function are defined based on the concept of distance.

8. The meaning and interpretation of the Intermediate Value Theorem become obvious when represented geometrically (Figure 4).

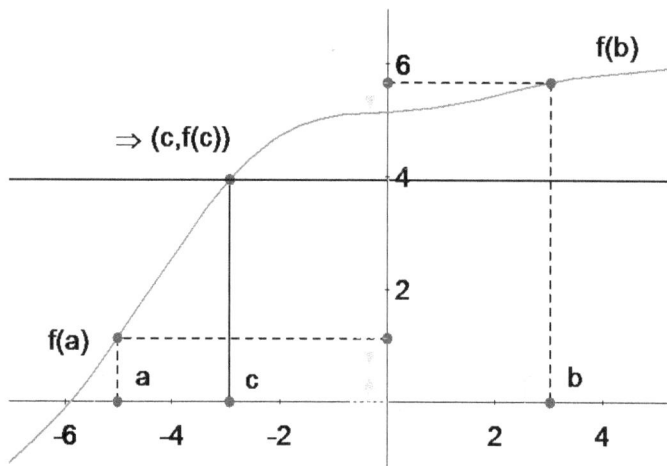

Figure 4. Intermediate Value Theorem

According to the NCTM Standards, "They [students] should be adept at visualizing, describing, and analyzing situations in mathematical terms. And they need to be able to

---

justify   and   prove   mathematically   based   ideas."
(http://standards.nctm.org/document/chapter7/index.htm). Certainly these standards are also
appropriate for calculus students. Many calculus problems can be visualized with the help of
geometric representation. Optimization problems that deal with areas and perimeters of
geometric shapes provide solutions to geometric problems. A geometric image helps in
setting up primary equations in related rates problems. In general, visual representation can
help the interpretation and solution of an analytic problem, sometimes providing a more
effective problem solving method.

The *Geometry Expressions* software allows for both dynamic visual representations and
symbolic algebraic output thus providing students the opportunity not only to visualize the
calculus problem but also to see in it in a more general way. This book offers interactive
calculus explorations that provide problems of different types:

1. Introduction of a concept (for example, the definition of a tangent line)

2. Proof of important theorems (for example, the Mean Value Theorem)

3. Calculation of geometric measures (for example, area)

4. Determination of mutual relationship between geometric objects (for example,
    parallel or perpendicular lines)

5. Computations (for example, finding limits or integrals)

Development of geometric intuition helps further the study of calculus. It supports organic
connections between geometry and calculus as it was historically developed based on
geometric ideas.

Special Features of *Geometry Expressions*

Dynamic Geometry software, such as *Geometer's Sketchpad* and *Cabri* has facilitated an
inductive approach to geometry. These technologies have been widely adopted in the last 20
years and a vast amount of creativity has been brought to bear on applying it to the
educational process. However, this technology is not without its shortcomings. First it is
construction-based; geometric configurations which are easy to state declaratively must be
expressed in terms of sequential constructions. Whereas this action may provide an
intriguing intellectual challenge, it may also be a significant distraction from the task at
hand. Secondly, this software is numeric only, and does not have a convenient way of
interacting with Computer Algebra Systems (CAS).

Geometry Expressions software developed by Saltire Software, Inc. is a dynamic, constraint-based symbolic geometry system that allows students to model the geometry to be expressed algebraically and then solved automatically in an algebra system. *Geometry Expressions* provides a mathematical tool that produces outputs in a symbolic form as algebraic expressions using the input parameters.

We know students love using technology in the classroom. Work with a user-friendly software package increases students' interest and motivation in learning mathematics, in particular, calculus. Using *Geometry Expressions* software also provides students with the opportunity to learn calculus from a modern approach, but at the same time very much tied to its roots.

Geometry Expressions enables more extensive calculus investigation than is possible in a traditional course of calculus. Open-ended explorations and investigations reinforce students' intellectual development. We believe students appreciate challenges and enjoy taking ownership in the problem solving process.

How to Use These Materials

Before discussing the use of these explorations in a classroom, we feel that it is necessary to insert a word of caution about how NOT to use them. The problems contained in this book are not intended to be a Calculus curriculum, nor can they be used as a series of exercises to introduce students to *Geometry Expressions.*

These materials immediately plunge the user into the *Geometry Expressions* software. The premise of this book is that this software can alter how students learn the concepts of calculus. The software allows students to visualize such major calculus concepts as the derivative and the integral in a fundamental geometric way facilitating a more complete understanding. The *Geometry Expressions* software (*Gx* as we call it) allows for both dynamic visual representations and symbolic algebraic output thus providing students the opportunity not only to visualize the calculus concept but also to see in it in a more general way.

So, the first step is to acquire a degree of fluency with *Gx*. Saltire Software, the developers and publishers of *Geometry Expressions*, has provided an excellent introductory tutorial on its web site: www.geometryexpressions.com → Explore → Using a Constraint Approach to Teaching Mathematics. This tutorial provides more than a "How To" manual. It provides an orientation to the underlying concept of this constraint-based software. Failure to grasp the nature of this constraint basis will likely be the source of much confusion and discouragement. You don't need to know everything but you do need a working knowledge of the principle features of the software.

Note to *Geometers Sketchpad* ® and *Cabri* ® users: There is a superficial resemblance between *Geometers Sketchpad* ® or *Cabri* ® and *Gx* which can be at first deceiving. It took some of the authors a while to fully understand and appreciate the differences.

Once basic skill is acquired with the software, what next?

❖ Selecting the explorations to use: This book is a collection of 29 explorations. The approach used in this book allows the teacher to direct and encourage student conjectures and observations of calculus topics to motivate them to think critically. The constraint-based software allows students to test their understanding at a higher level by considering complex situations that require multiple level thinking. These explorations were developed with two goals in mind. Firstly, the calculus explorations are geometrically based and contain representative samples of many calculus techniques and methods. Secondly, the explorations display the different ways *Gx* can be used. Again, an effort has been made to include as wide a variety of uses as possible within this collection. A Cross Reference Table is provided to help you identify particular explorations that meet these goals.

The Cross Reference Table suggests a couple of practical approaches that a teacher could take to incorporate some of these explorations into the existing curriculum.

❖ Select explorations in order to review and deepen student understanding. For example, after having taught a unit on limits, a teacher might plan to use explorations from Chapter 1. Limits. Thus a teacher might decide to use the Exploration 1.2 where students explore the limit of $\frac{\sin x}{x}$ with the Squeeze Theorem or Exploration 1.3 where students discover that the perimeter of a regular n-gon converges to the circumference of a circle.

➢ Use the problems to deepen and extend your curriculum. Some of the problems use mathematical content that is not commonly found in American secondary schools calculus curriculum. For example, Explorations 2.3 and 2.4 where students discover properties of derivations of even and odd function, or Exploration 3.1 that introduces the concept of the envelope of a parabola, deal with the content that is not in most of our curricula. Yet these explorations extend the students understanding of these most fundamental ideas of the calculus curriculum.

➤ Select explorations in order to support and enhance AP calculus curriculum. The order of the explorations follows a typical AP Calculus AB to BC syllabus. The references to the AP calculus topics are provided in the Cross-Reference Table.

❖ Using the problems in the classroom: There is no one best way to use these problems. The differences from one classroom to the next are too varied to admit generalization. We do offer, however, some ideas for you to consider.

➤ The explorations are intended to be used as teacher-led activities. The questions posed in the explorations are thoughtfully selected to develop reasoning skills and deeper understanding of calculus concepts. This has to be done in class with the teacher facilitating discussion.

➤ As students become more familiar with the software, the teacher can take a much less directed role in the discovery process. Let students freely explore with *Gx*. However, the teacher should continue the whole class discussions of the questions posed in each exploration.

➤ Ask students whether their answers to questions or solutions of the problems are reasonable. Always encourage students to articulate their thinking and their results and to justify their work. Be alert for spontaneous student conjectures that can be explored.

➤ You don't have to finish the exploration by the end of the class period. It is okay to let students ponder it for a day or two before you bring up the exploration again.

➤ Assign a problem to be done by one group of students using traditional calculus techniques and possibly the graphing calculator and by another group using *Gx*. Exploration 3.5 *"Floating Log"* could be especially fruitful to explore using the two different techniques.

➤ Share the problems with teachers and students beyond the calculus classroom. Exploration 3.6 "The Art Gallery Problem" could provoke an interesting discussion with students and teachers in a geometry class.

This set of explorations provides samples of how calculus reasoning and logical thinking could be developed using *Gx*. In order to develop mastery, complement these explorations with a significant number of practice problems where students can apply the ideas they learned through the explorations in this book. In the end, your own professional judgment will be the best guide to incorporating these rich explorations into your classes.

CROSS REFERENCE TABLE

Problem Name	Pp.	Pre-requisite knowledge	Key concept	Level	Class Time	AP Calculus AB and BC* Topic
1.1) The Formal Definition of a Limit	22	Properties of absolute value Concavity of monotonic functions	ε-δ definition of limit	3	60 – 90 min.	*Limits of functions (including one-sided limits)
1.2) The Squeeze Theorem	34	Basic trigonometric ratios in a unit circle Inverse trigonometric functions Area of a circular sector Limit of a function at a point	$\lim\limits_{x \to 0} \dfrac{\sin x}{x} = 1$	1	45 min.	*Limits of functions (including one-sided limits)
1.3) The Area of a Circle	43	The area of a regular polygon inscribed in a circle with given radius A bounded increasing sequence converges $\lim\limits_{x \to 0} \dfrac{\sin(ax)}{x} = a$	Limits at infinity Horizontal asymptote	1	45 min.	Asymptotic and unbounded behavior
2.1) Exploring Tangent Lines	54	The slopes of tangent and normal lines to the curve are opposite reciprocals. Point-slope equation of a line Derivative formulas for basic functions	Equation of a tangent line Slope of the tangent line is equal to the derivative of the function	1	45 min.	Derivative at a point

Problem Name	Pp.	Pre-requisite knowledge	Key concept	Level	Class Time	AP Calculus AB and BC* Topic
2.2) The Mean Value Theorem	61	Slope of a line given two points The derivative as a slope of a tangent line	The Mean Value Theorem	1	45 min.	Derivative as a function
2.3) Derivative of Even Functions	70	Symmetry of graphs of even and odd functions Basic derivatives rules Derivative as a slope of a tangent line $\text{sign}(x) = \begin{cases} \dfrac{x}{\lvert x \rvert}, x \neq 0 \\ 1, x = 0 \end{cases}$	If $f(-x) = f(x)$ then $f'(-x) = -f'(x)$	2	45 – 60 min.	Derivative as a function
2.4) Derivative of Odd Functions	78	Symmetry of graphs of even and odd functions Basic derivatives rules Derivative as a slope of a tangent line $\text{sign}(x) = \begin{cases} \dfrac{x}{\lvert x \rvert}, x \neq 0 \\ 1, x = 0 \end{cases}$	If $f(-x) = -f(x)$ then $f'(-x) = f'(x)$	2	45 – 60 min.	Derivative as a function

Problem Name	Pp	Pre-requisite knowledge	Key concept	Level	Class Time	AP Calculus AB and BC* Topic
2.5) Differentiability of a Piece-Wise Function	86	Graph of a piecewise function Continuity of a function Limit of a function at a point Derivative as the slope of the tangent	Differentiability of a piecewise function at the joint point	3	60 – 90 minutes	Concept of a derivative
2.6) Derivative of an Inverse Function	98	Graphical and symbolic representations of inverse functions Existence of the inverse function Reflection over the identity line $y = x$ Slope of a tangent line to a function as the derivative of the function	The relationship between the derivative of a function and the derivative of its inverse	2	45 – 60 minutes	Applications of derivatives
3.1) Envelope of a Parabola	110	Quadratic formula Slope-intercept form of the equation of a line Derivative as the slope of the tangent	Concavity of a function	1	45 minutes	Applications of derivatives
3.2) Linear Approximation	117	Differentiability of a function Derivative as the slope of the tangent Point-slope form of an equation of a line	Using a tangent line to approximate the value of a function at a point	2	45 – 60 minutes	Derivative at a point

Problem Name	Pp.	Pre-requisite knowledge	Key concept	Level	Class Time	AP Calculus AB and BC* Topic
3.3) Newton's Method	130	Point-slope form of the equation of a line Recursive formula Derivative as a slope of the tangent line Convergence and divergence of sequences Differentiability and continuity of functions	Use tangent lines to approximate the zeros of a function	2	45 – 60 min.	Applications of derivatives
3.4) Rectangle in a Semicircle	143	Pythagorean Theorem Product rule and chain rule of differentiation 1st and 2nd derivative tests	Optimization	1	45 min.	Applications of derivatives
3.5) Floating Log	150	Pythagorean Theorem Right triangle trigonometry Similar triangle theorems Derivatives of trigonometric functions The chain rule 1st and 2nd derivative tests	Optimization	2	45 – 60 min.	Applications of derivatives

Problem Name	Pp.	Pre-requisite knowledge	Key concept	Level	Class Time	AP Calculus AB and BC* Topic
3.6) Art Gallery	161	Inverse trigonometric relationships Derivatives of trigonometric functions Product, quotient, and chain rules of differentiation 1st derivative test	Optimization	2	45 – 60 min.	Applications of derivatives
4.1) Representation of the Antiderivative	170	Derivative as a slope of the tangent Translation Basic rules for differentiation	Algebraic and graphical representation of the antiderivative	2	45 – 60 min.	Techniques of antidifferentia-tion
4.2) Fundamental Theorem of Calculus	181	Antiderivatives of basic functions	Definite integral	1	45 min.	Fundamental Theorem of Calculus
4.3) Second Fundamental Theorem of Calculus	190	Derivative as a slope of the tangent Basic rules of differentiation Antiderivative Definite integral The Fundamental Theorem of Calculus	Definite integral with a variable upper limit Area function	2	45 – 60 min.	Fundamental Theorem of Calculus
4.4) Integral of an Inverse Function	197	Inverse functions Reflection Integrals of basic functions	Integration	2	45 – 60 min.	Techniques of antidifferentia-tion

Problem Name	Pp.	Pre-requisite knowledge	Key concept	Level	Class Time	AP Calculus AB and BC* Topic
4.5) The Trapezoid Method	208	Area of trapezoid Sigma notation	Area under a curve	2	45 – 60 min.	Numerical approximations to definite integrals
4.6) Minimum Area	219	Derivatives of basic functions The 1st and 2nd derivative tests The area between curves Definite integral	Optimiza-tion	1	45 min.	Applications of integrals
5.1) Orthogonal Trajectory to a Circle	228	Derivative as a slope of the tangent line The angle between curves The slopes of perpendicular lines Separation of variables for solving 1st order linear differential equations.	Separation of variables	2	45 – 60 min.	Applications of antidifferentiation
5.2) Orthogonal Trajectory to a Hyperbola	236	Derivative as a slope of the tangent line The angle between curves The slopes of perpendicular lines Separation of variables for solving 1st order linear differential equations	Separation of variables	2	45 – 60 min.	Applications of antidifferentiation

Problem Name	Pp.	Pre-requisite knowledge	Key concept	Level	Class Time	AP Calculus AB and BC* Topic
6.1) Infinite Stairs	246	Limit at infinity	Constant series	1	45 min.	Concept of a series.
6.2) The Snowman Problem	258	Area of a circle. Sigma notation. Sum of a geometric series.	Geometric series	1	45 min.	Series of constants
6.3) Trigonometric Delight	267	Similar triangles Right triangle trigonometry relationships Sum of geometric series	Geometric series	2	45 – 60 min.	Series of constants
6.4) Converging or Diverging?	280	Improper integral Sigma notation Partial sum of a series Integral test of series convergence	Telescopic series Harmonic series	3	60 – 90 min.	Series of constants
7.1) Folium of Descartes Using Parametric Equations	292	Parametric equations Asymptotic behavior Limit at a point Limit at infinity	Asymptotic behavior and symmetry of a parametric curve	2	45 – 60 min.	*Parametric, polar, and vector functions
7.2) Folium of Descartes Using Polar Coordinates	302	Polar coordinates and polar equations Parametric derivative Derivative of trigonometric functions 1st derivative test	Horizontal and vertical tangent lines at a pole of a polar curve. Local maximum for a polar curve.	2	45 – 60 min.	*Parametric, polar, and vector functions

AP Calculus Topic

Each problem is categorized according to AP Calculus AB and BC curriculum topics found on
http://apcentral.collegeboard.com/apc/public/courses/descriptions/index.html.

Pre-requisite knowledge

Here we have listed the properties, formulas or definitions used in the activity beyond what might be expected in an activity for a particular topic.

Level of difficulty:

Level 1 is characterized by explorations that can be accomplished

- with straightforward calculus techniques
- within one class period

Level 2 is characterized by explorations that may require

- a deeper understanding of calculus concepts and techniques
- more than one class period to complete

Level 3 is characterized by explorations that require

- a deeper understanding of abstract concepts and formal definitions of calculus
- more than one class period to complete.

Class Time

Our suggestions for how much class time to allot are based on the assumption that students are

- familiar with the pre-requisite calculus content along with the other mathematics involved (*e.g.* algebra, precalculus, geometry), and
- familiar enough with *Geometry Expressions* to progress through the steps following the teacher's guidance.

---

1. Limits

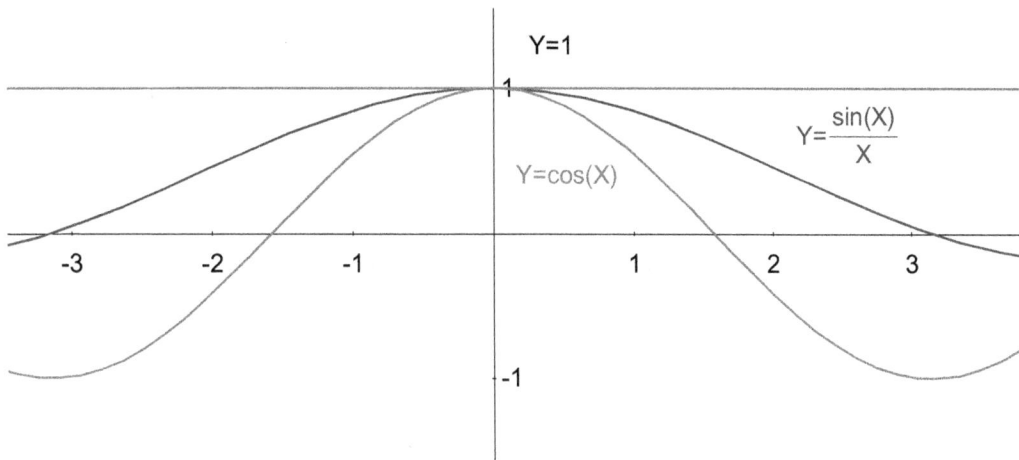

1.1 The Formal Definition of a Limit

<u>Exploration 1.1</u>: In this activity you will explore the formal definition of the limit graphically in order to understand its meaning.

The definition of limit: Let f be a function defined on an open interval containing c (except possibly at c) and let L be a real number. The statement $\lim_{x \to c} f(x) = L$ means that for each $\varepsilon > 0$ there exists $\delta > 0$ such that if $0 < |x - c| < \delta$, then $|f(x) - L| < \varepsilon$.

In this problem we will assume that $f(x)$ is a strictly monotonic function on an open interval containing point c and we will use the following notations:

$$y_1 = f(c) - \varepsilon = f(x_1), \quad y_2 = f(c) + \varepsilon = f(x_2).$$

SUMMARY

<u>Mathematics Objectives:</u>

- Visualize the formal $\delta - \varepsilon$ definition of the limit of a function at a point.

- Explore and determine a general method of finding δ given ε for a strictly monotonic function

<u>Vocabulary:</u>

- Concavity

- Monotonic function

- Limit of a function at a point

<u>Pre-requisites:</u>

- Properties of absolute value

- Concavity of monotonic functions

<u>Problem Notes:</u>

- The $\delta - \varepsilon$ definition of the limit is a difficult concept for most students to understand. Visualizing the problem with specific functions leads to better understanding of the concept in general. Furthermore, this problem helps students' understanding of the definition of limit by using a simple graphical illustration.

- Students first explore the limit of function $f(x) = x^2$ at point $x = c$. They construct an interval on the y-axis such that $|f(x) - f(c)| \le \varepsilon$. Then they use the graph of the function to find the δ-interval on the x-axis such that $|x - c| \le \delta$ for all corresponding points. While determining the method of finding δ, students have to take into account the fact that the function is concave up.

- Students then explore the limit of function $f(x) = \sqrt{x}$ at $x = c > 0$ using the same approach, but this time in order to find δ they have to take into account that the function is concave down.

- At the end of the exploration students are asked to generalize their findings for any monotonic concave up or concave down function.

Technology skills:

- Draw: functions, points, line segments, and circles

- Constrain: point proportional along the curve, perpendicular lines, radius of a circle

- Calculate: coordinates of a point

Extensions:

Given $f(x) = x^2$ at $x = 0$. For an arbitrarily small $\varepsilon > 0$ find δ such that if $|x| < \delta$ then $|f(x)| < \varepsilon$.

STEP-BY-STEP INSTRUCTIONS

LIMIT OF $f(x) = x^2$ AT POINT $x = c$

1. Plot a function $f(x) = x^2$ with a point on the function at $x = c$.

   a. Use **Toggle grid and axes** to show the axes without the grid.

   b. Choose **Draw** → Function. Choose Cartesian for the Type of function, and enter the expression for the function in Y=.

---

c. Plot a point on the curve by choosing **Draw** → Point. Label point C then select it and the graph of the function, and choose **Constrain** → Point proportional along curve. Type c in the open edit box. Select point C and choose **Calculate** → Coordinates.

2. Choose $\varepsilon > 0$ and graphically find an interval $[y_1,\ y_2]$ on the y-axis, such that $|f(x) - f(c)| = |f(x) - c^2| \le \varepsilon$.

Note: since our software can only operate on closed intervals, from now on we will use non strict inequalities to define the intervals.

a. Choose **Draw** → Line Segment and plot a segment from point C to the y - axis. Select the segment and the y – axis and choose **Constrain** → Perpendicular.

b. Select the point of intersection with the y-axis and choose **Calculate** → Coordinates.

c. Choose **Draw** → Circle, and plot circle with the center at the point *(0, f(c))*. Select the circle and choose **Constrain** → Radius. Type ε in the open edit box.

d. Plot the points of intersection of the circle and the y-axis. These points will define the interval $|f(x) - c^2| \le \varepsilon$.

e. Select each point you found, one at a time, and choose **Calculate** → Coordinates.

Q1. What are the x-coordinates of the points on the graph of $y = x^2$ that correspond to the end points of the interval $|f(x) - c^2| \le \varepsilon$?

A1: At the point y_1, the corresponding point on the function is defined by: $x_1^2 = c^2 - \varepsilon \Rightarrow x_1 = \sqrt{c^2 - \varepsilon}$, and at the point y_2, , the corresponding point on the function is defined by: $x_2^2 = c^2 + \varepsilon \Rightarrow x_2 = \sqrt{c^2 + \varepsilon}$. (We find the positive values of x_1 and x_2, assuming $\varepsilon \le c^2$)

3. Plot the two points on the graph of $y = x^2$ with the coordinates you found.

a. Choose **Draw** → Points and plot two points on the graph of *f(x)*, one above and one below the point C.

b. Select the point you constructed above point C and the graph of *f(x)*, and choose **Constrain** → Proportional. In the open edit box type sqrt(c^2+ε).

---

c. Select the point you constructed below point C and the graph of $f(x)$, and choose **Constrain** → Proportional. In the open edit box type sqrt(c^2-ε).

d. Connect the points on the graph with the corresponding points on the y-axis by using **Draw** → Line Segment.

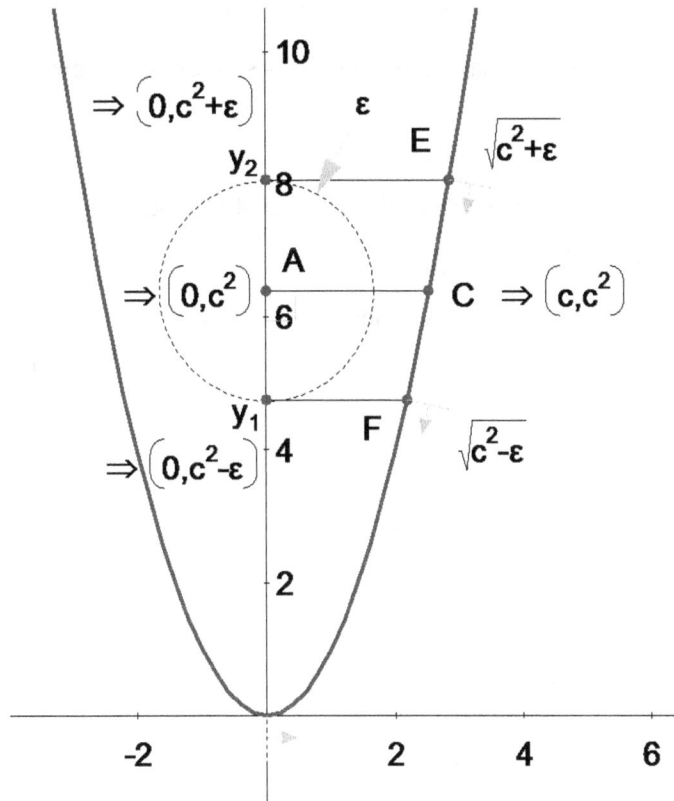

Q2. Can you find a value of δ, so that if $|x-c| \leq \delta$, then $|f(x)-c^2| \leq \varepsilon$? Explain this from a geometric point of view.

A2: We need to plot points on the x-axis with coordinates x_1 and x_2, and choose the smaller of the two distances, $c - x_1$ and $x_2 - c$. Then the largest possible value of δ is the smaller of the two distances.

4. Plot an interval $|x-c| \leq \delta$ on the x-axis such that $|f(x)-c^2| < \varepsilon$ and determine δ geometrically.

a. Choose **Draw** → Line Segment and plot a segment from point C to x - axis. Select the segment and the x – axis and choose **Constrain** → Perpendicular. Label the point c.

b. In a similar manner, Choose **Draw** → Segment and plot perpendiculars to the x-axis from the other two points on the graph of the function.

c. In a similar manner to what we did on the y-axis, draw a circle with center at c and radius δ by choosing **Draw** → Circle. Select the circle and choose **Constrain** → Radius, and type δ for the radius.

d. Plot the points of intersection of the circle and the x-axis. These points will define the interval $|x - c| \leq \delta$.

e. Change δ using a slider in the Variables panel to meet the condition in question 2.

Q3. How do you choose δ for this particular function? Explain.

A3: *Choose δ as the smaller of two values: $x_2 - c$ or $c - x_1$. For this particular case, since the function is concave up, $c - x_1 > x_2 - c$, so $\delta = x_2 - c$.*

Q4. For the given function and the given value of c is it possible to find a δ for any given small ε? What does it mean?

A4: *For any (sufficiently small) ε we can find corresponding points on the x-axis, and thus, we will be able to find δ. That means the function has a limit at point c.*

*Note: Students can select ε in the **Variables** panel and use a slider to change its value to observe the effect on the interval of values of x. Then, they can adjust the value of δ.*

5. Change the value of c, choose an ε, and find δ graphically to confirm your answers.

   a. In the Variables panel, click on the variable c and set its value using a slider or typing the value.

Note: If you do not see the point C on the graph of f(x) after you entered the value of c, it is most likely out of the visible window. Adjust the value of c to have the point within viewing window or use Zoom Out.

   b. Click on the variable ε and set its value using a slider or typing the value.

   c. Click on the variable δ. Adjust its upper and lower values if needed, and use a slider to modify the radius of the circle until you find the value of δ such that the interval containing all values of x around c is enclosed in the circle.

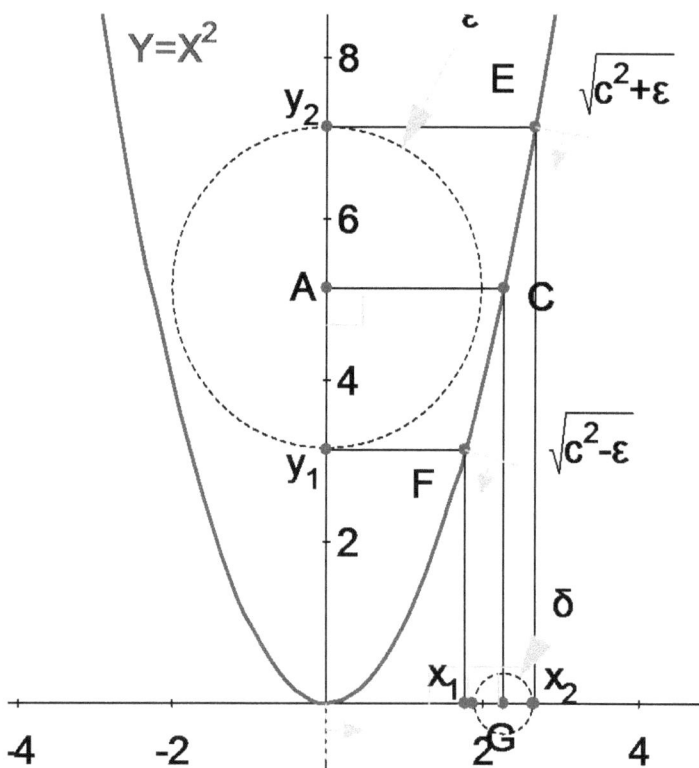

Q5. What is the equation for δ for the function $y = x^2$ in order to prove that this function has a limit at the point $x = c$ on the domain of the function?

A5: Since the function is concave up, the right interval will always be smaller than the left interval for $x > 0$, and thus $\delta = \sqrt{c^2 + \varepsilon} - c$. For $x < 0$ due to symmetry, the smaller interval is the left one, so $\delta = \sqrt{c^2 + \varepsilon} + c$. Generalizing, $\delta = \sqrt{c^2 + \varepsilon} - |c|$ for all points in the domain of the function.

LIMIT OF $y = \sqrt{x}$ AT THE POINT $x = c > 0$

Q6. Choose $\varepsilon > 0$. What are the x-coordinates of points on the graph of $y = \sqrt{x}$ that correspond to the end points of the interval $\left| f(x) - \sqrt{c} \right| \le \varepsilon$?

A6: At the point y_1, $\sqrt{x_1} = \sqrt{c} - \varepsilon \Rightarrow x_1 = \left(\sqrt{c} - \varepsilon \right)^2$, and at the point y_2, $\sqrt{x_2} = \sqrt{c} + \varepsilon \Rightarrow x_2 = \left(\sqrt{c} + \varepsilon \right)^2$. (Assuming that $\varepsilon \le \sqrt{c}$).

6. Change the function and constraints to explore limits of the function $y = \sqrt{x}$.

   a. Double-click on $Y = X^2$ and change that to sqrt(X).

   b. Double click on the Point proportional constraint for the points on the curve below and above point C. For the point with coordinates (x_1, y_1) enter (sqrt(c)-ε)^2. For the point with coordinates (x_2, y_2) enter (sqrt(c)+ε)^2.

   c. Adjust the values of c and ε.

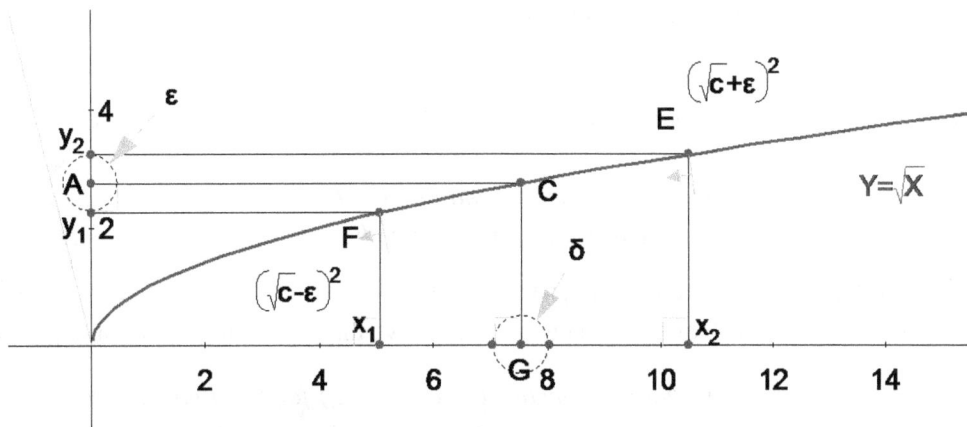

Q7. What is the value of δ in case of the function $y = \sqrt{x}$ at $x = c$? Explore with the software and explain.

A7: *Based on the work above, students should be able to state that $\delta = c - (\sqrt{c} - \varepsilon)^2$ or $\delta = (\sqrt{c} + \varepsilon)^2 - c$, whichever is smaller. Using a slider to change the value of δ shows that the right interval is larger than the left, and thus $\delta = c - (\sqrt{c} - \varepsilon)^2$. This is due to the fact that the function is concave down.*

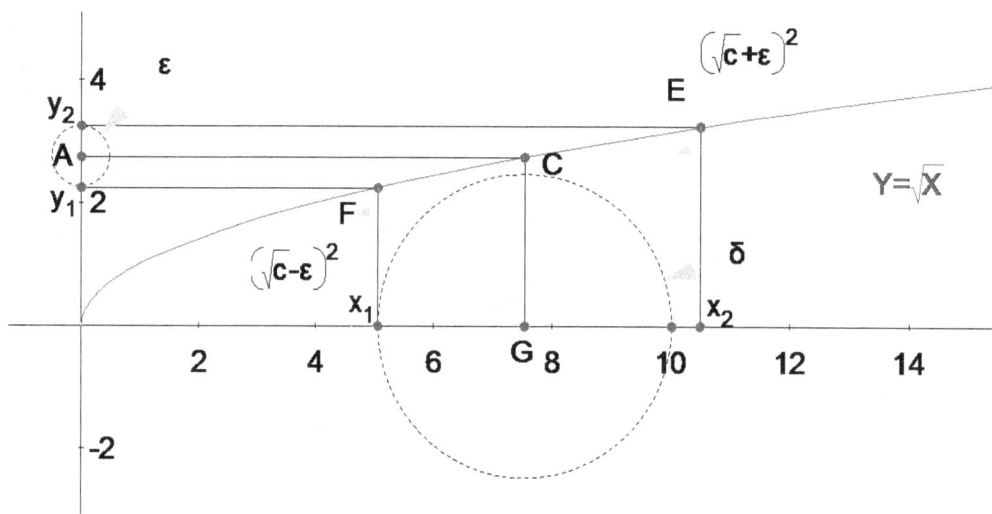

Q8. Consider a general continuously increasing function (around point c except, possibly at point c) $y = f(x)$. How do you choose δ to prove that the function has a limit at point c?

A8: *For functions that are concave up, $\delta = x_2 - c$; for functions that are concave down, $\delta = c - x_1$. This is due to the fact that increments of the function are equal, and thus increments of independent variables depend on how fast the function is increasing. If the concavity changes on the interval containing c, then the general rule is not possible.*

Q9. Consider a general continuously decreasing function (around point c except, possibly, at point c) $y = f(x)$. How do you choose δ to prove that function has a limit at point c?

A9: *Similarly, for functions that are concave up, $\delta = c - x_1$; for functions that are concave down, $\delta = x_2 - c$. This is due to the fact that increments of the function are equal, and thus increments of independent variables depend on how fast the function is decreasing.*

If the concavity changes on the interval containing c, then the general rule is not possible.

The Formal Definition of a Limit

<u>Exploration 1.1:</u> In this activity you will explore the formal definition of the limit graphically in order to understand its meaning.

The definition of limit: Let f be a function defined on an open interval containing c (except possibly at c) and let L be a real number. The statement $\lim_{x \to c} f(x) = L$ means that for each $\varepsilon > 0$ there exists $\delta > 0$ such that if $0 < |x - c| < \delta$, then $|f(x) - L| < \varepsilon$.

In this problem we will assume that $f(x)$ is a strictly monotonic function on an open interval containing point c and we will use the following notations:

$$y_1 = f(c) - \varepsilon = f(x_1), \quad y_2 = f(c) + \varepsilon = f(x_2).$$

LIMIT OF $f(x) = x^2$ AT POINT $x = c$

1. Plot a function $f(x) = x^2$ with a point on the function at $x = c$.

2. Choose $\varepsilon > 0$ and graphically find an interval $[y_1, y_2]$ on the y-axis, such that
    $$|f(x) - f(c)| = |f(x) - c^2| \le \varepsilon.$$

Q1. What are the x-coordinates of the points on the graph of $y = x^2$ that correspond to the end points of the interval $|f(x) - c^2| \le \varepsilon$?

3. Plot the two points on the graph of $y = x^2$ with the coordinates you found.

Q2. Can you find a δ, so that if $|x - c| \le \delta$, then $|f(x) - c^2| \le \varepsilon$? Explain this from a geometric point of view.

4. Plot an interval $|x - c| \le \delta$ on the x-axis such that $|f(x) - c^2| < \varepsilon$ and determine δ geometrically.

Q3. How do you choose δ for this particular function? Explain.

Q4. For the given function and the given value of c is it possible to find a δ for any given small ε? What does it mean?

5. Change the value of c, choose an ε, and find δ graphically to confirm your answers.

Q5. What is the equation for δ for the function $y = x^2$ in order to prove that this function has a limit at a point $x = c$ on the domain of the function?

LIMIT OF $y = \sqrt{x}$ AT POINT $x = c > 0$

Q6. Choose $\varepsilon > 0$. What are the x-coordinates of points on the graph of $y = \sqrt{x}$ that correspond to the end points of the interval $\left| f(x) - \sqrt{c} \right| \leq \varepsilon$?

6. Change the function and constraints to explore limits of the function $y = \sqrt{x}$.

Q7. What is the value of δ in case of the function $y = \sqrt{x}$ at $x = c$? Explore with the software and explain.

Q8. Consider a general continuously increasing function (around point c except, possibly at point c) $y = f(x)$. How do you choose δ to prove that function has a limit at point c?

Q9. Consider a general continuously decreasing function (around point c except, possibly, at point c) $y = f(x)$. How do you choose δ to prove that function has a limit at point c?

1.2 The Squeeze Theorem

Exploration 1.2: Using geometric and algebraic approaches, explore and find the $\lim\limits_{x \to 0} \dfrac{\sin x}{x}$.

SUMMARY

Mathematics Objectives:

- Explore geometric representation of the Squeeze Theorem as a method of finding the special limit, $\lim\limits_{x \to 0} \dfrac{\sin x}{x}$.

- Explore algebraic representation of the Squeeze Theorem as a method of finding the special limit, $\lim\limits_{x \to 0} \dfrac{\sin x}{x}$.

Vocabulary:

- Unit circle

- Circular sector

Pre-requisites:

- Basic trigonometric ratios in a unit circle

- Inverse trigonometric functions

- Area of a circular sector

- Limit of a function at a point

Problem Notes:

- The Squeeze Theorem is often glossed over in a traditional calculus curriculum, but is necessary to fully understand several important derivative proofs.

- In this exploration, students first establish that the area of a circular sector is bounded by the areas of two triangles and thus arrive at inequality $\sin x < x < \tan x$.

- Students then use this relationship to investigate the $\lim\limits_{x \to 0} \dfrac{\sin x}{x}$ graphically.

Technology skills:

- Draw: circle, arc, line segment, polygon, function

- Construct: polygon interior

- Constrain: angle, radius, point incident to a segment, perpendicular lines

- Calculate: area

Extension:

Find $\lim\limits_{x \to 0} \dfrac{\arcsin(x)}{x}$ using the Squeeze Theorem.

STEP-BY-STEP INSTRUCTIONS

GEOMETRIC EXPLORATION

1. ⓘ Check your preferences.

    a. In the window on the right of the status bar to make sure you are working in **Radians**.

    b. In the menu **Edit→** Preferences, choose the **Math** tab, **Output** window, click Use Assumptions and set the value to True.

2. Draw a unit circle with its center (point A) at the origin.

    a. Use **Toggle grid and axes** to show the axes without the grid.

    b. Choose **Draw →** Circle and plot it with the center at the origin. Select the circle and choose **Constrain →** Radius. In the open edit box type 1.

3. Draw a triangle with vertices A(0,0), C(1,0), and a vertex D on the circle in the 1st quadrant. Set the measure of angle ∠CAD = x.

    a. Choose **Draw →** Polygon. Click first on the circle center (point A), then on the circle at $x = 1$ (point C), and then on the circle in an arbitrary point in the 1st quadrant (point D).

    b. Select segments AC and AD and choose **Constrain →** Angle. Type x in the open edit box.

Q1. What is the area of this triangle ACD?

A1: $A_1 = \frac{1}{2}r^2 \sin x = \frac{\sin x}{2}$, *since r = 1.*

3.   Verify your calculations using the software.

    a.   Select triangle interior and choose **Calculate** → Area.

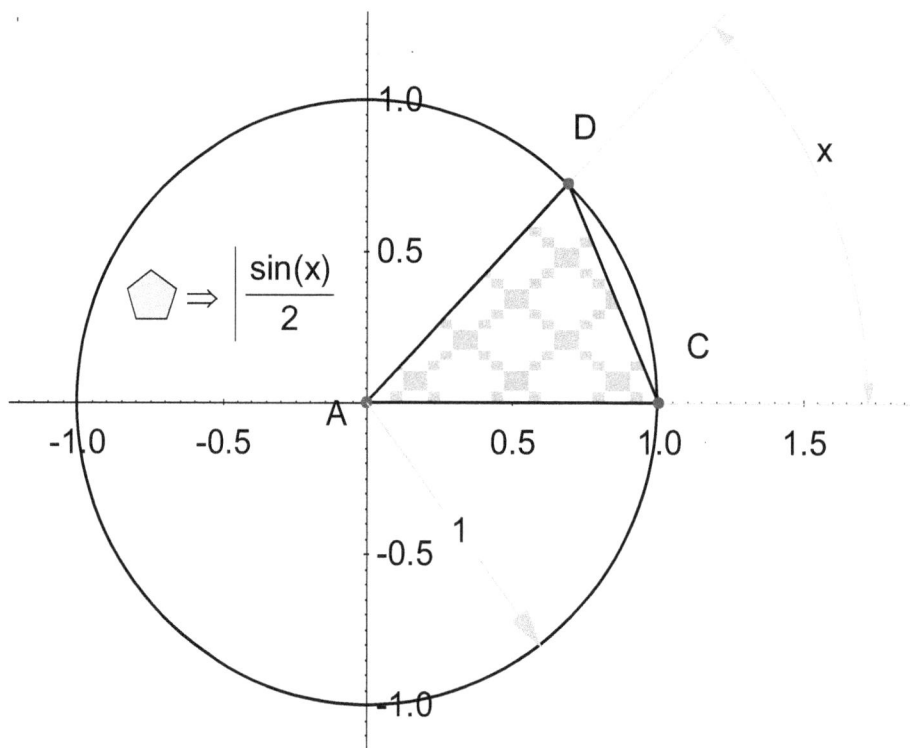

4.   Construct a circular sector with arc CD, sides AC and AD.

    a.   Hide the triangle to avoid confusion (select inside the triangle and the segment CD, right click and select **Hide** from the context menu).

    b.   Choose **Draw** → Arc and click the endpoints C and D.

    c.   Select the arc CD and segments AC and AD, and choose **Construct** → Polygon

Q2.   What is the area of this circular sector?

A2: $A_2 = \frac{1}{2}r^2 x = \frac{x}{2}$, *since r = 1.*

5. Verify your calculations using the software.

    a. Select the sector interior and choose **Calculate** → Area.

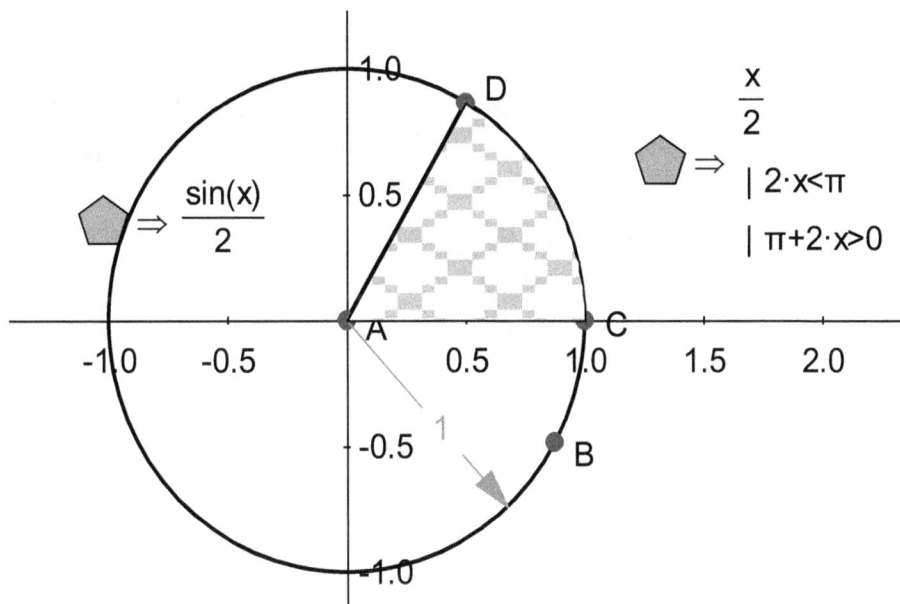

6. Construct a right triangle ACE, where CE is tangent to the circle at point C and segment AE contains point D.

    a. Choose **Draw** → Segment and draw segments AE and CE.

    b. Select segment AE and point D, and choose **Constrain** → Incident.

    c. Select segments AC and CE, and choose **Constrain** → Perpendicular.

    d. Select segments AC, AE, and CE and choose **Construct** → Polygon.

Q3. What is the area of this right triangle?

A3: $A_3 = \dfrac{1}{2}r^2 \tan x = \dfrac{\tan x}{2}$, *since r = 1*

7. Verify your calculations using the software.

    a. Select polygon interior and choose **Calculate** → Area.

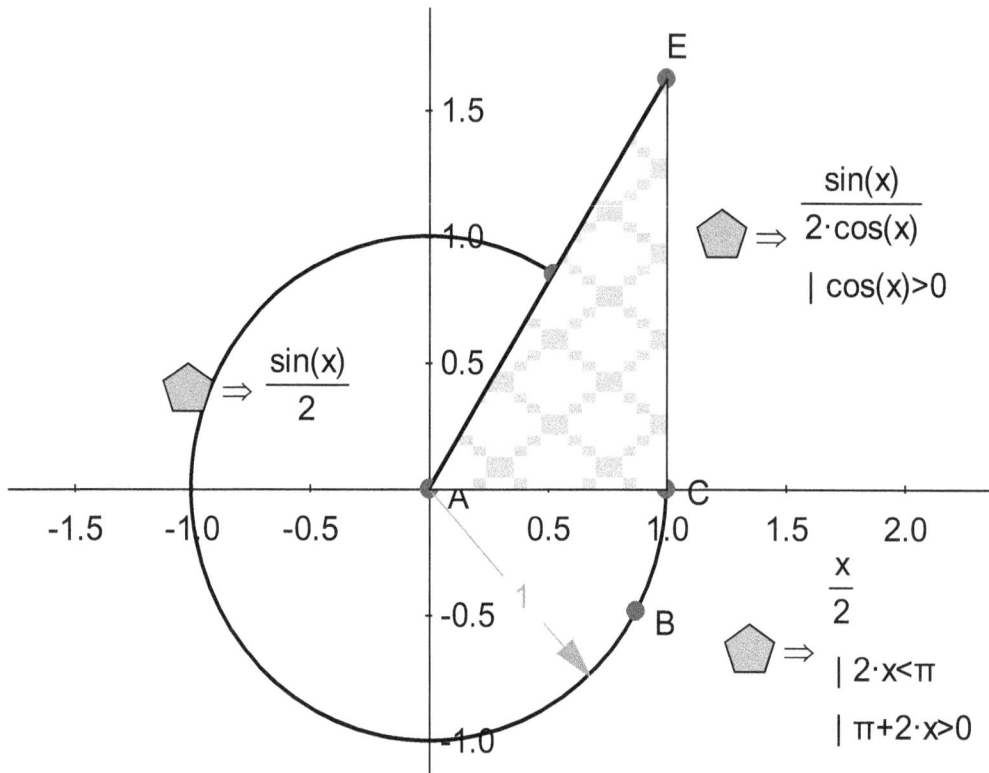

Q4. Compare the area of the circular sector ACD with areas of triangles ACD and ACE and write an inequality for these areas. Express this inequality in terms of x

A4: $A_1 < A_2 < A_3 \Leftrightarrow \sin x < x < \tan x$.

INVESTIGATING $\lim\limits_{x \to 0} \dfrac{\sin x}{x}$

Q5. Using your answer to question 4, re-write the inequality for $\dfrac{\sin x}{x}$.

A5: *All functions in the inequality are positive for values of x in the 1st quadrant. Consider the reciprocal to the inequality:* $\dfrac{1}{\tan x} < \dfrac{1}{x} < \dfrac{1}{\sin x}$. *Multiply the inequality by $\sin x$, we get:* $\dfrac{\sin x}{\tan x} < \dfrac{\sin x}{x} < \dfrac{\sin x}{\sin x} \Leftrightarrow \cos x < \dfrac{\sin x}{x} < 1$.

8.   Open a new document and plot the graph of $f(x) = \dfrac{\sin x}{x}$ and functions that bound $f(x)$ from above and from below based on your answer to question 5.

   a.   Choose **Draw** → Function. Choose Cartesian type and in the Y= prompt enter expression for $f(x)$.

   b.   Choose **Draw** → Function. Choose Cartesian type and in the Y= prompt enter the value of the upper bound function.

   c.   Choose **Draw** → Function. Choose Cartesian type and in the Y= prompt enter the value of the lower bound function.

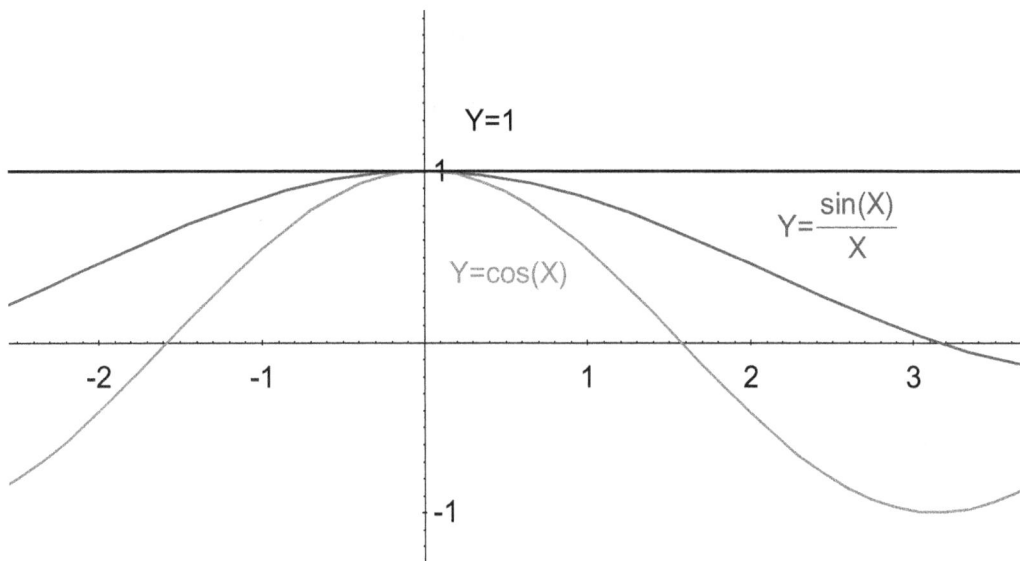

Q6.   Does the limit of $f(x) = \dfrac{\sin x}{x}$ as $x \to 0$ exist? If so, what is the limit? Justify your answer.

A6:

   1.   *the inequality* $\cos x < \dfrac{\sin x}{x} < 1$ *is valid for all nonzero x on an open interval* $\left(-\dfrac{\pi}{2}, \dfrac{\pi}{2}\right)$ *containing $x=0$ due to the fact that all functions in the inequality are even;*

   2. $\lim\limits_{x \to 0} \cos x = 1;$

3. $\lim\limits_{x \to 0} 1 = 1$. *By the Squeeze Theorem, then* $\lim\limits_{x \to 0} \dfrac{\sin x}{x} = 1$.

The Squeeze Theorem

Exploration 1.2: Using geometric and algebraic approaches, explore and find the $\lim\limits_{x \to 0} \dfrac{\sin x}{x}$.

GEOMETRIC EXPLORATION

1. Draw a unit circle with its center (point A) at the origin.

2. Draw a triangle with vertices A(0,0), C(1,0), and a vertex D on the circle in the 1st quadrant. Set the measure of angle $\angle CAD = x$.

Q1. What is the area of the triangle ACD?

3. Verify your calculations using the software.

4. Construct a circular sector with arc CD, sides AC and AD.

Q2. What is the area of this circular sector?

5. Verify your calculations using the software.

6. Construct a right triangle ACE, where CE is tangent to the circle at point C and segment AE contains point D.

Q3. What is the area of this right triangle?

7. Verify your calculations using the software.

Q4. Compare the area of the circular sector ACD with the areas of triangle ACD and ACE and write an inequality for the areas. Express this inequality in terms of x

INVESTIGATING $\lim\limits_{x \to 0} \dfrac{\sin x}{x}$

Q5. Using your answer to question 4, re-write inequality for $\dfrac{\sin x}{x}$.

8. Open a new document and plot the graph of $f(x) = \dfrac{\sin x}{x}$ and functions that bound $f(x)$ from above and from below based on your answer to question 5.

Q6. Does the limit of $f(x) = \dfrac{\sin x}{x}$ as $x \to 0$ exist? If so, what is the limit? Justify your answer.

1.3 Area of a Circle

Exploration 1.3: Given a circle with radius r, a regular n-sided polygon is inscribed in the circle. If n increases without bound, will the sequence of areas of the inscribed polygons have a limit? If so, what is this limit and what does it mean geometrically?

SUMMARY

Mathematics Objectives:

- Explore the sequence of the areas of regular polygons inscribed in a circle.

- Explore the limit of this sequence as the number of polygon sides increases without bound.

- Determine that the limit of this sequence exists and is equal to the area of the circle.

Vocabulary:

- Bounded sequence

- Monotonic sequence

- Converging sequence

- Horizontal asymptote

Pre-requisites:

- The area of a regular polygon inscribed in a circle with a given radius

- The Double Angle Formula

- A bounded increasing sequence converges

- $\lim\limits_{x \to 0} \dfrac{\sin(ax)}{x} = a$

Problem Notes:

1. This problem is an example of how limits can be used to find the formula for the area of a circle. By applying the tools of calculus we let A_n be an area of a regular n-sided inscribed polygon and S be the area of a circle and if $A_n \leq S$ for all n, where S is the smallest upper bound for the sequence A_n, and if $\lim\limits_{n \to \infty} A_n = L$, then $S = L$.

2. Students first explore the area of a regular polygon inscribed in a circle of given radius. They develop formulas for a triangle, square, and general n-sided polygon, and verify their formulas with the help of the software.

3. Students then explore the behavior of the sequence formed by the areas of polygons as n (the number of sides) increases without bound. Intuitively and visually, students should arrive at the understanding that the sequence is increasing. For algebraic justification, it is suggested to analyze the ratio of the areas A_n and A_{2n}.

4. Students then graph the area of the polygon function and observe the asymptotic behavior of this function, with the horizontal asymptote being the area of the circle. Thus they visualize the convergence of the area of the inscribed polygon to the area of the circle in a different way.

Technology skills:

- Draw: circle, N-gon, function

- Construct: polygon

- Constrain: radius

- Calculate: area

Extension:

Explore the convergence of the sequence of the areas of the regular polygons circumscribed about the circle of a given radius as the number of sides of the polygon increases without bound.

STEP-BY-STEP INSTRUCTIONS

EXPLORING THE AREA OF INSCRIBED POLYGONS

1. Draw a circle and constrain its radius to be r.

   a. Use **Toggle grid and axes** to hide the axes and the grid.

   b. Choose **Draw** → Circle. Select the circle and choose **Constrain** → Radius. In the open edit box type r.

2. Draw an equilateral triangle inscribed in the circle.

   a. Choose **Draw** → N-gon. Click first on the circle center and then on the circle.

---

b.   In the open edit box type 3 for the number of sides.

Q1. What is the area of the inscribed equilateral triangle?

A1: $A_3 = 3\left(\dfrac{1}{2}r^2 \sin\dfrac{2\pi}{3}\right) = \dfrac{3\sqrt{3}}{4}r^2$

3.   Verify your calculations using the software.

Select the triangle interior and choose **Calculate** → Area.

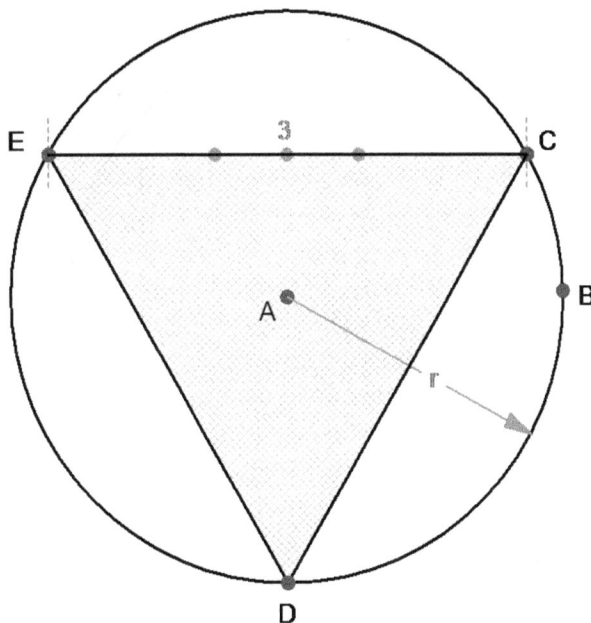

4.   Open a new document.

5.   As you did before, draw a circle and constrain its radius to be *r*.

a.   Use **Toggle grid and axes** to hide the axes and the grid.

b.   Choose **Draw** → Circle. Select the circle and choose **Constrain** → Radius. In the open edit box type *r*.

6.   Draw a square inscribed into the circle.

a. Choose **Draw** → N-gon. Click first on the circle center and then on the circle.

b. In the open edit box type 4 for the number of sides.

Q2. What is the area of the inscribed square?

A2: $A_4 = 4\left(\dfrac{1}{2}r^2 \sin\dfrac{2\pi}{4}\right) = 2r^2$

7. Verify your calculations using the software.

Select the interior of the square and choose **Calculate** → Area.

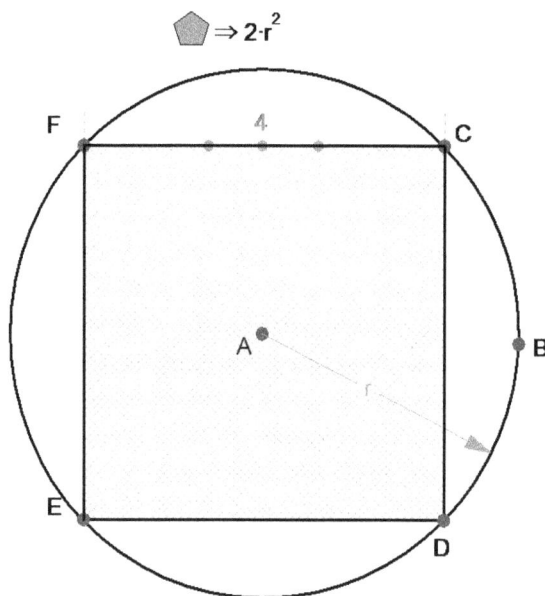

8. Open a new document.

9. As you did before, draw a circle and constrain its radius to be *r*.

a. Use **Toggle grid and axes** to hide the axes and the grid.

b. Choose **Draw** → Circle. Select the circle and choose **Constrain** → Radius. In the open edit box type *r*.

10. Draw a regular *n*-gon inscribed in the circle.

a. Choose **Draw** → N-gon. Click first on the circle center and then on the circle.

b.   In the open edit box type *n* for the number of sides.

Q3.   What is the area of the inscribed regular *n*-sided polygon?

A3: $A_n = n\left(\dfrac{1}{2}r^2 \sin\dfrac{2\pi}{n}\right) = \dfrac{nr^2}{2}\sin\dfrac{2\pi}{n}$

11.   Verify your calculations using the software.

Select the polygon interior and choose **Calculate** → Area.

$$\pentagon \Rightarrow \left|\frac{n\cdot r^2 \cdot \sin\left(\dfrac{2\cdot\pi}{n}\right)}{2}\right|$$

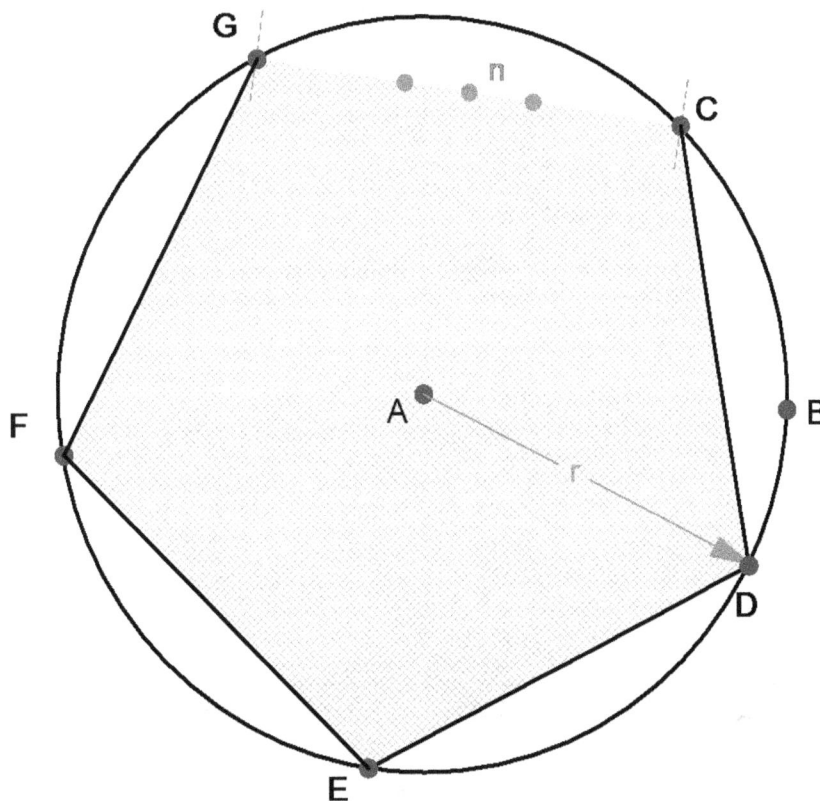

INVESTIGATING A SEQUENCE OF POLYGONAL AREAS

Q4. What happens to the sequence of polygon areas, A_n, when we double the number of sides n? Explain.

A4: *As n increases, the polygon area increases. In order to demonstrate that the sequence A_n is increasing, consider A_{2n}/A_n :*

$$\frac{A_{2n}}{A_n} = \frac{2nr^2 \sin\frac{2\pi}{2n}}{nr^2 \sin\frac{2\pi}{n}} = \frac{2\sin\frac{\pi}{n}}{\sin\frac{2\pi}{n}} = \frac{2\sin\frac{\pi}{n}}{2\sin\frac{\pi}{n}\cos\frac{\pi}{n}} = \frac{1}{\cos\frac{\pi}{n}} > 1, \text{ for } n \geq 3 \text{ since } \cos\frac{\pi}{n} \text{ is}$$

increasing as n increases.

Q5. Does the sequence, A_n, have an upper bound? Explain.

A5: *For all values of n,* $A_n = \frac{nr^2}{2}\sin\frac{2\pi}{n} \leq \frac{nr^2}{2}\cdot\frac{2\pi}{n} \leq \pi r^2$

12. Plot the graph of the function that contains all terms of the sequence A_n and the upper bound if it exists on the same graph to confirm your answers to questions 5 and 6.

   a. Choose **Draw** → Function. Choose Cartesian type and in the Y= prompt enter your expression for A_n replacing n with x.

   b. Choose **Draw** → Function. Choose Cartesian type and in the Y= prompt enter the value of the upper bound.

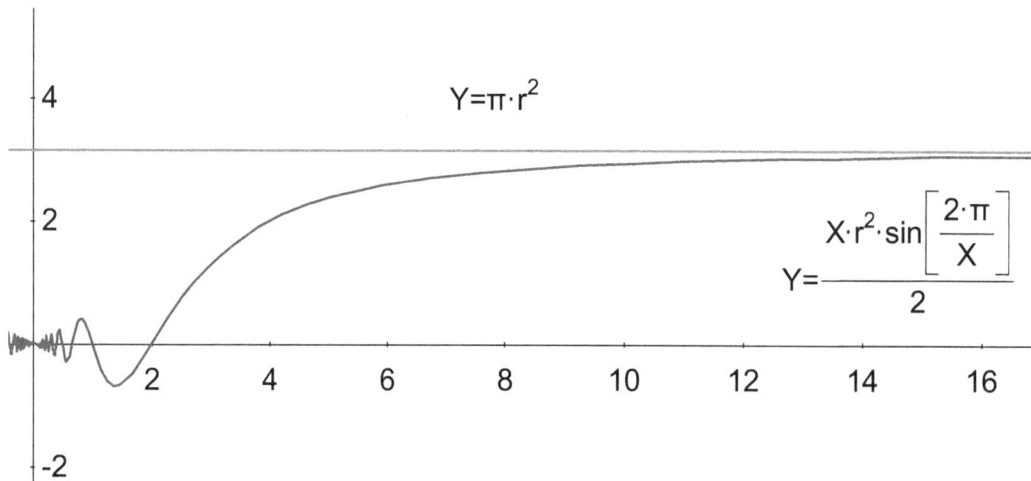

Q6. Does the sequence of areas of the regular inscribed polygon have a limit when n increases without a bound? What is the limit?

A6: *The limit exists since sequence A_n is increasing and has an upper bound. Here is the calculation for the limit:* $\lim\limits_{n\to\infty} A_{ni} = \lim\limits_{n\to\infty}\left(\dfrac{nr^2}{2}\sin\left(\dfrac{2\pi}{n}\right)\right).$ *Substitute* $x = \dfrac{2\pi}{n}$ *then*

$n = \dfrac{2\pi}{x}$. *Since* $n \to \infty$, *then* $x \to 0$, *and the limit becomes*

$\lim\limits_{n\to\infty} A_{ni} = \pi r^2 \lim\limits_{x\to 0}\left(\dfrac{\sin x}{x}\right) = \pi r^2.$

Q7: What is the meaning of this limit from geometric point of view?

A7: *The limit of the polygon area is equal to the area of the circle.*

Area of a Circle

Exploration 1.3: Given a circle with radius r. A regular n-sided polygon is inscribed in the circle. If n increases without bound, will the sequence of areas of the inscribed polygons have a limit? If so, what is this limit and what does it mean geometrically?

EXPLORING AREA OF INSCRIBED POLYGONS

1. Draw a circle and constrain its radius to be r.

2. Draw an equilateral triangle inscribed into the circle.

Q1. What is the area of the inscribed equilateral triangle?

3. Verify your calculations using the software.

4. Open a new document.

5. Draw a circle and constrain its radius to be r.

6. Draw a square inscribed into the circle.

Q2. What is the area of the inscribed square?

7. Verify your calculations using the software.

8. Open a new document.

9. Draw a circle and constrain its radius to be r

10. Draw a regular polygon inscribed into the circle.

Q3. What is the area of the inscribed regular n-sided polygon?

11. Verify your calculations using the software.

INVESTIGATING A SEQUENCE OF POLYGON AREAS

Q4. What happens to the sequence of polygon areas, A_n, when we increase the number of sides n? Explain. (Hint: compare A_n and A_{2n}).

Q5. Does the sequence, A_n, have an upper bound? Explain.

12. Plot the graph of the function that contains all the terms of the sequence A_n and the upper bound if it exists on the same graph to confirm your answers to questions 5 and 6.

Q6. Does the sequence of areas of the regular inscribed polygon have a limit when n increases without bound? What is the limit?

Q7: What is the meaning of this limit from a geometric point of view?

2. Derivatives

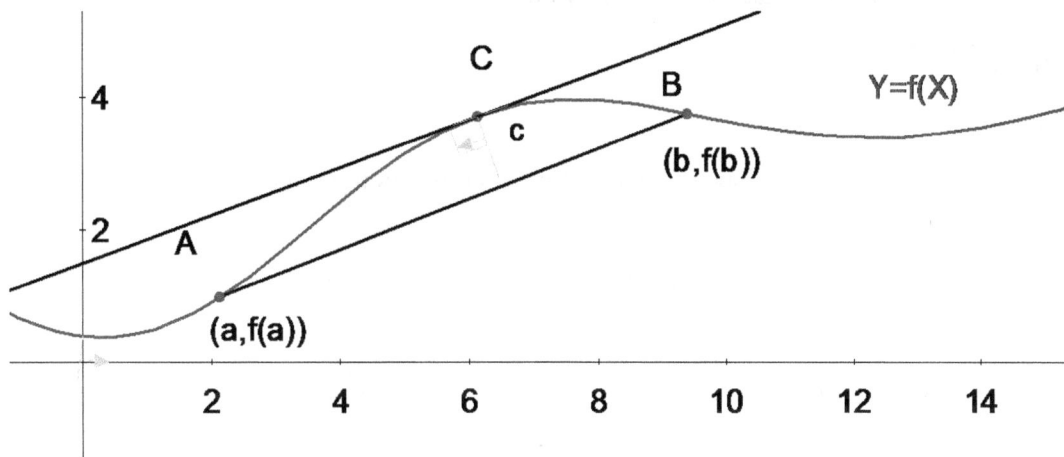

2.1 Exploring Tangent Lines

<u>Exploration 2.1:</u> Given a differentiable function $y = f(x)$. A tangent line to the function is drawn through the point A(a, $f(a)$). The equation of the tangent line can be written in the form $y = m(x - x_1) + y_1$. What is the meaning of m and (x_1, y_1) in the equation and how are these related to the function $f(x)$ and the coordinates of the point A?

a) First, consider the special case $f(x) = x^2$

b) And then generalize for an arbitrary differentiable function $f(x)$.

SUMMARY

<u>Mathematics Objectives:</u>

- Discover that the slope of a tangent line to a function $y = f(x)$ at the point $(a, f(a))$ is equal to $f'(a)$.

- Determine that the equation of the tangent line at $(a, f(a))$ is $y = f'(a)(x - a) + f(a)$.

<u>Pre-requisites:</u>

- The slopes of tangent and normal lines to the curve are opposite reciprocals.

- Point-slope equation of a line.

- Derivative formulas for basic functions.

<u>Problem Notes:</u>

- Students explore the relationship between the slope of a tangent line and the derivative of a function at a given point on the graph of the function. First they use a specific function such as $y = x^2$. *Geometry Expressions* provides students with an expression for the slope of the tangent line that they compare with the derivative of the function at the same point. Students derive the equation of the tangent line and verify it with the help of the software.

- Then students generalize their results to a generic function $y = f(x)$. With the help of the software students establish the definition of the tangent line, as the line through a point on a curve that has the slope equal to the derivative of the function at this point.

- If students have difficulty deriving the equation for the tangent line in the first part of the problem, we suggest that they explore several different specific functions prior to moving to the general case of $y = f(x)$. Consider $y = x^n$ for integer values of n, $y = \sqrt{x}$, $y = \cos x$ and $y = \sin x$.

Technology skills:

- Draw: function

- Construct: tangent to a curve

- Constrain: point proportional along a curve

- Calculate: coordinates of a point, slope of a line, an implicit equation

Extension:

A normal line is constructed on the graph of a function $y = f(x)$ at a point $x = a$. Find the relationship between the slope of the normal and the derivative of the function at this point.

STEP-BY-STEP INSTRUCTIONS

INVESTIGATION OF $Y = X^2$

1. Use **Toggle grid and axes** to show the axes without a grid.

2. Choose **Draw** → Function. Select Type → Cartesian and enter x^2 to plot the parabola $Y = x^2$.

3. Select the plot of the function and choose **Construct** → Tangent. The tangent line and the point of tangency (point A) will be displayed.

4. Select point A and the parabola and choose **Constrain** → Point proportional along curve. Type a into the edit box. This constrains the x-coordinate of point A to be a.

*Note: although we constrained $x = a$, a is a parameter that can change its numeric value. Students can either drag point A along the curve or vary the value of a in the **Variables** panel.*

Q1. What are coordinates of point A? Explain how you found these values.

A1: Since $x = a$ and point A is on the curve $y = x^2$, its coordinates are (a, a^2)

---

5. Calculate the coordinates of point A using software.

    a. Select the point A.

    b. Choose **Calculate** → Coordinates.

Q2. What do you think is the slope of the tangent line to *f(x)* through point A?

A2: Students' answers may vary.

6. Calculate the slope of the tangent line using the software.

    a. Select the tangent line.

    b. Choose **Calculate** → Slope.

Q3. Do you see any relationship between the slope of the tangent line and the coordinates of the point of tangency?

A3: The slope of tangent line is 2a and the coordinates of point A are (a, a^2). Since $\frac{dy}{dx} = (x^2)' = 2x$, then $2a = 2x|_{x=a} = f'(a)$. Thus, the slope of the tangent line is equal to the value of the derivative of the function at the given point.

7. Find the equation for the tangent line using the software.

    a. Select the tangent line.

    b. Choose **Calculate** → Implicit equation.

Q4. In this equation what is the slope and what is the y-intercept? Explain your results.

A4: The equation of the tangent line is $y = 2ax - a^2$. The slope is 2a and y-intercept is $-a^2$

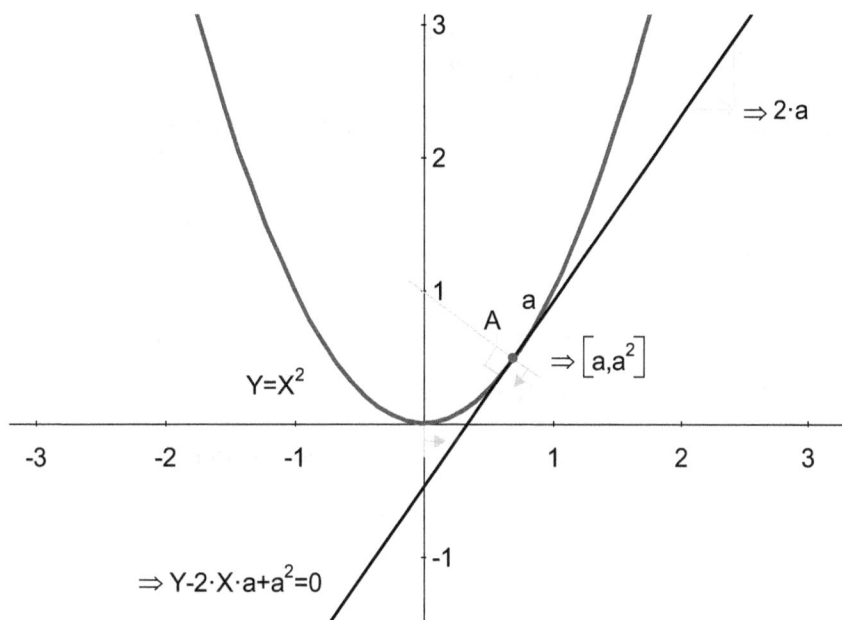

Q5. Describe the steps you can use to determine the equation of a tangent line to a parabola $y = x^2$ at $x = a$.

A5: *One of the possible answers could be the following:*

1. *Find the derivative of a function at a given point – it is the slope of the tangent line so m=2a.*

2. *In the point-slope form of the equation of a line, use the coordinates of point A:* $y - a^2 = 2a(x - a) \Leftrightarrow y = 2ax - a^2$.

INVESTIGATION OF *Y = F(X)*

Note: You can have your students use other concrete examples of functions (for example, y = x^3, y = sin x) and ask students to apply their method of finding the equation of the tangent line. To verify their calculations with Geometry Expressions students only need to edit the expression for the function by double-clicking the function and typing the new expression. The calculations will be automatically updated. Then they can transition to the more general case y = f(x).

Q6. Can you use the same steps to find an equation of a tangent line for a general differentiable function $y = f(x)$ at $x = a$?

A6: Students should recognize that the steps to find an equation of the tangent line are the same for all functions.

Q7. What is the slope of the tangent line to the function $f(x)$ at $x = a$?

A7: $m = f'(a)$

Q8. What is an equation of the tangent line to the function $f(x)$ at $x = a$?

A8: $y - f(a) = f'(a)(x - a)$ or $y = f'(a)(x - a) + f(a)$

8. In order to verify your answers, double-click on the equation of the parabola and type $f(x)$ instead of x^2.

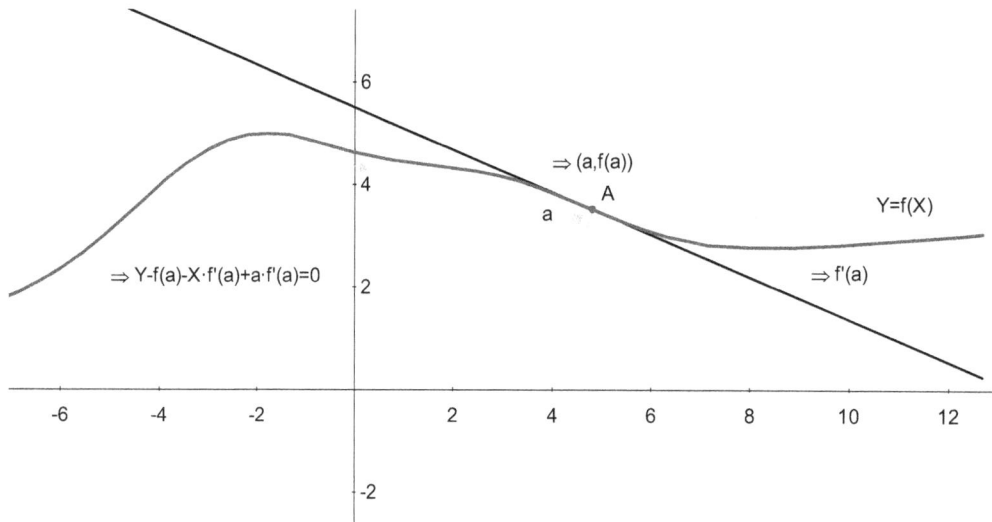

Exploring Tangent Lines

Exploration 2.1: Given a differentiable function $y = f(x)$. A tangent line to the function is drawn through the point $A(a, f(a))$. The equation of the tangent line can be written in the form $y = m(x - x_1) + y_1$. What is the meaning of m and (x_1, y_1) in the equation and how are these related to the function $f(x)$ and the coordinates of the point A?

    a) First, consider the special case $f(x) = x^2$

    b) And then generalize for an arbitrary differentiable function $f(x)$.

INVESTIGATION OF $Y = X^2$

1.   Use **Toggle grid and axes** to show the axes without a grid.

2.   Choose **Draw** → Function. Select Type → Cartesian and enter x^2 to plot the parabola $Y = x^2$.

3.   Select the plot of the function and choose **Construct** → Tangent. The tangent line and point of tangency will be displayed (point A).

4.   Select point A and parabola and choose **Constrain** → Point proportional along curve. Type a in the edit box to constrain the x-coordinate of point A to be a.

Q1. What are the coordinates of point A? Explain how you found those values.

5.   Calculate the coordinates of point A using the software.

Q2. What is the slope of the tangent line?

6.   Calculate the slope of the tangent line using Gx.

Q3. Do you see any relationship between the slope of the tangent line and the coordinates of the point of tangency?

7.   Find an equation of the tangent line using Gx.

Q4. In this equation what is the slope and what is the y-intercept? Explain your results in terms of your answers to Q7.

Q5. Describe the steps you can use to determine the equation of the tangent line to the parabola $y = x^2$ at $x = a$.

INVESTIGATION OF $Y = F(X)$

Q6. Can you use the same steps to find an equation of a tangent line for a general differentiable function $y = f(x)$ at $x = a$? Explain how you would do so.

Q7. What is the slope of the tangent line to the function $f(x)$ at $x = a$?

Q8. What is an equation of the tangent line to the function $f(x)$ at $x = a$?

8. In order to verify your answers double-click on the equation of the parabola and type $f(x)$ instead of x^2. The calculations will change automatically.

2.2 Mean Value Theorem

Exploration 2.2: Explore and explain the Mean Value Theorem: if $f(x)$ is defined and continuous on the interval $[a, b]$ and differentiable on (a, b), then there is at least one number c in the interval (a, b) (that is $a < c < b$) such that $f'(c) = \dfrac{f(b) - f(a)}{b - a}$.

SUMMARY

Mathematics Objectives:

- Discover the Mean Value Theorem for a generic continuous and differentiable function $y = f(x)$

- Apply the Mean Value Theorem to specific functions and solve for c.

Pre-requisites:

- Slope of a line given two points

- Derivative as a slope of a tangent line

Problem Notes:

- The Mean Value Theorem has a fundamental meaning in calculus. Sometimes this theorem is also called the Lagrange Theorem.

- Students first explore the general statement of the theorem by dragging the tangent line along the graph of a function $y = f(x)$ on a given interval and confirming the existence of a point where the tangent line is parallel to the secant line through the endpoints of the interval.

- Students then apply the theorem to three specific functions and solve for values of c where the tangent line is parallel to the secant line.

Technology skills:

- Draw: function, line segment, tangent line to a curve

- Constrain: coordinates of a point, a point proportional along a curve

- Calculate: slope

Extension:

Given an ellipse with a chord, prove that there exist two tangents that are parallel to the chord.

STEP-BY-STEP INSTRUCTIONS

GENERAL STATEMENT

1. Draw a function $y = f(x)$. Modify the shape of the function as needed.

   a. Use **Toggle grid and axes** to show the axes without grid.

   b. Choose **Draw** → Function. Select Type → Cartesian. In the Y = prompt type f(x).

   c. In order to modify the shape of the function:

      • Click on the curve, hold the mouse and drag the curve. A small circle and the name of the parameter will appear on the screen.

      • Click in another part of the function and drag it.

      • There are total of 5 different parameters that are responsible for the shape of the function, and you can change each one of them by dragging the function in 5 different places.

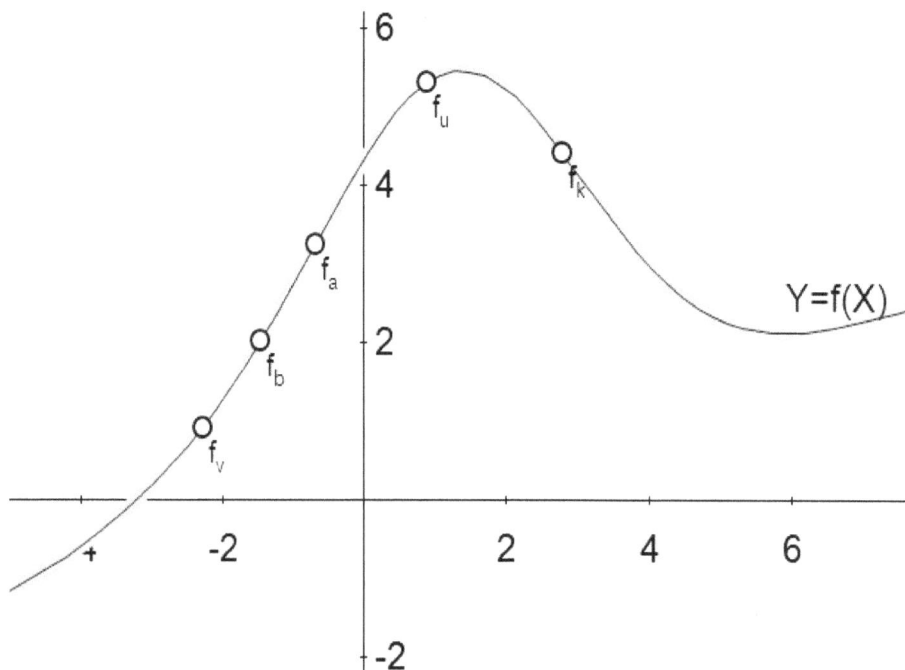

2. Draw chord AB to the curve of the function and constrain the coordinates of the endpoints as A(*a*, *f*(*a*)) and B(*b*, *f*(*b*)), so that *a* < *b*.

   a. Choose **Draw** → Segment and draw an arbitrary segment AB with the endpoints *not* on the curve.

   b. Select point A and choose **Constrain** → Coordinates. In the open edit box type (a, f(a)).

   c. Select point B and choose **Constrain** → Coordinates. In the open edit box type *(b, f(b))*.

   d. Move either point to satisfy the condition *a* < *b*.

3. Draw a tangent line to the graph of the function at an arbitrary point C (*c*, *f*(*c*)), so that *a* < *c* < *b*.

   a. Select the plot of the function and choose **Construct** → Tangent. The tangent line and point of tangency will be displayed (point C).

   b. Select point C and the graph of *f(x)*, and choose **Constrain** → Point proportional along curve. Type *c* in the open edit box.

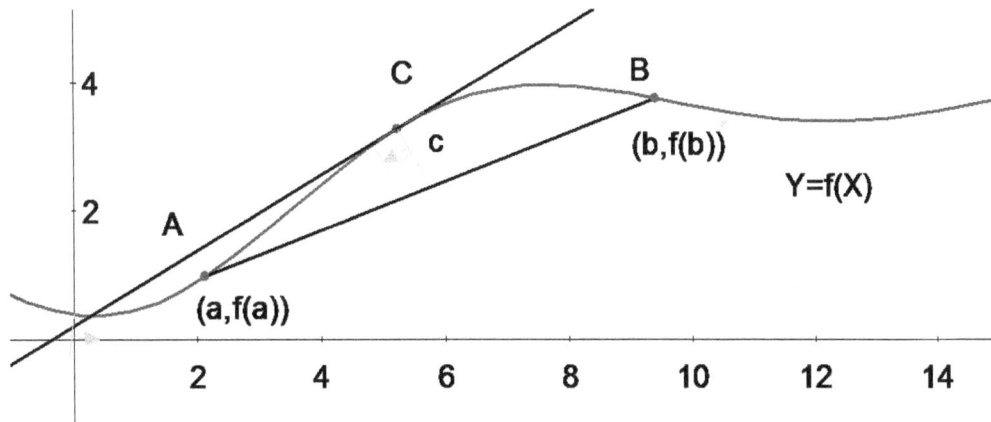

Q1. What is the slope of the chord AB?

A1: $\dfrac{f(b)-f(a)}{b-a}$

3. Verify your answer using *Gx*.

   a. Select the chord and choose **Calculate** → Slope.

Q2. What is the slope of the tangent line through point C?

A2: $f'(c)$

4. Verify your answer using *Gx*.

   a. Select the tangent line and choose **Calculate** → Slope.

Q3. Is it possible to find a location of the point C on the graph of *f*(*x*) between points A and B so that the slope of tangent line is equal to the slope of the chord AB?

A3: Yes, it is possible when the chord and the tangent lines are parallel. Students can drag point C along the curve to see this case.

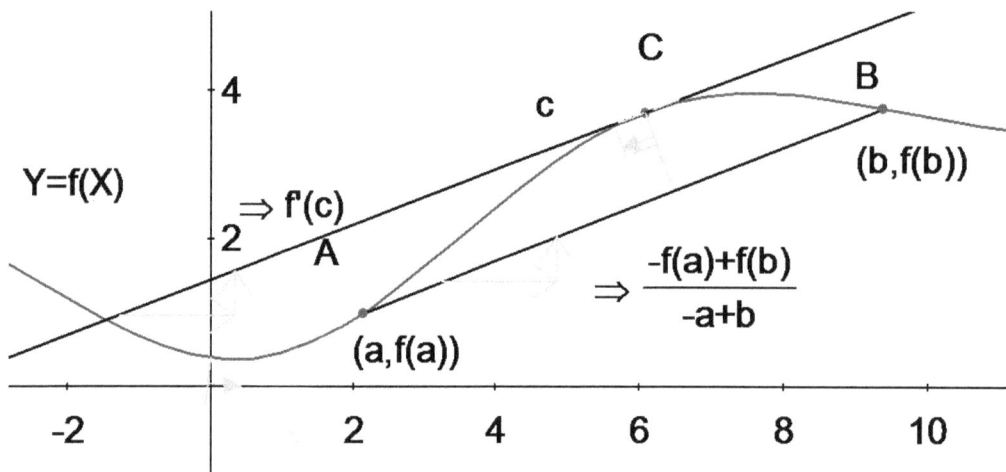

Q4. Is it possible to find more than one location of the point C that satisfies this condition? (Reminder: [*a*, *b*] is an arbitrary interval such that *f*(*x*) is continuous on [*a*, *b*] and differentiable on (*a*, *b*))

A4: It can be possible depending on the function and the interval.

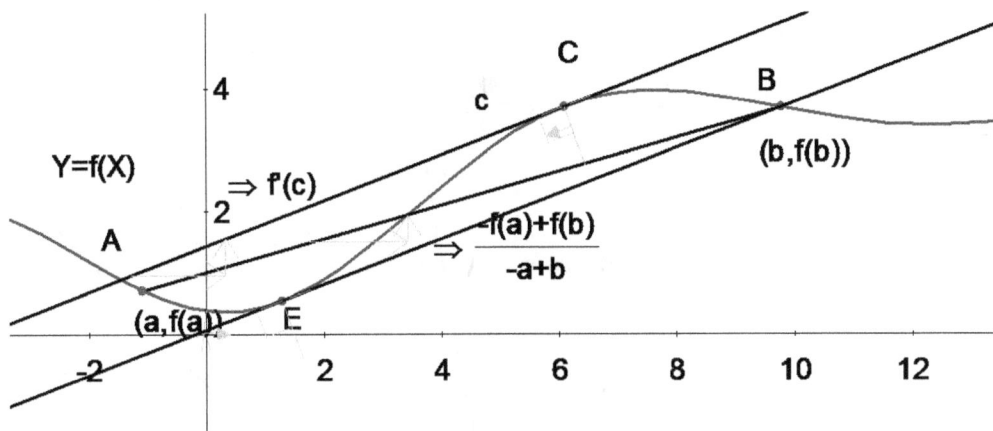

Q5. What would be the relative position of the chord to the tangent line(s) in this case?

A5: The tangent line is parallel to the chord.

APPLICATIONS

Q6. For the functions and intervals provided below use both analytical and graphical methods to

1. Graph the given function on the given interval.

2. Find and graph the secant line connecting the endpoints of the given interval.

3. Find all points on the given interval where the function satisfies the Mean Value Theorem and graph the corresponding tangent lines.

a. $y = x^2$ on $[-1, 2]$; b. $y = \sqrt{x}$ on $[1, 9]$; c. $y = x - 2\sin x$ on $[-\pi, \pi]$;

Note: For each problem students can use the construction they created for the general function and input the specific information:

1. Move point C into the given interval for the problem,

2. Type the specific function in place of f(x),

3. Type the new coordinates for a and b, e.g. (-1,1) and (2,4)

4. Students then calculate the value(s) of c and replace the variable c with this calculated value confirming that the lines are parallel and that c does exist.

Solutions:

a. $y = x^2$, on [-1, 2], $2c = \dfrac{4-1}{3} = 1$, *so* $c = \dfrac{1}{2}$.

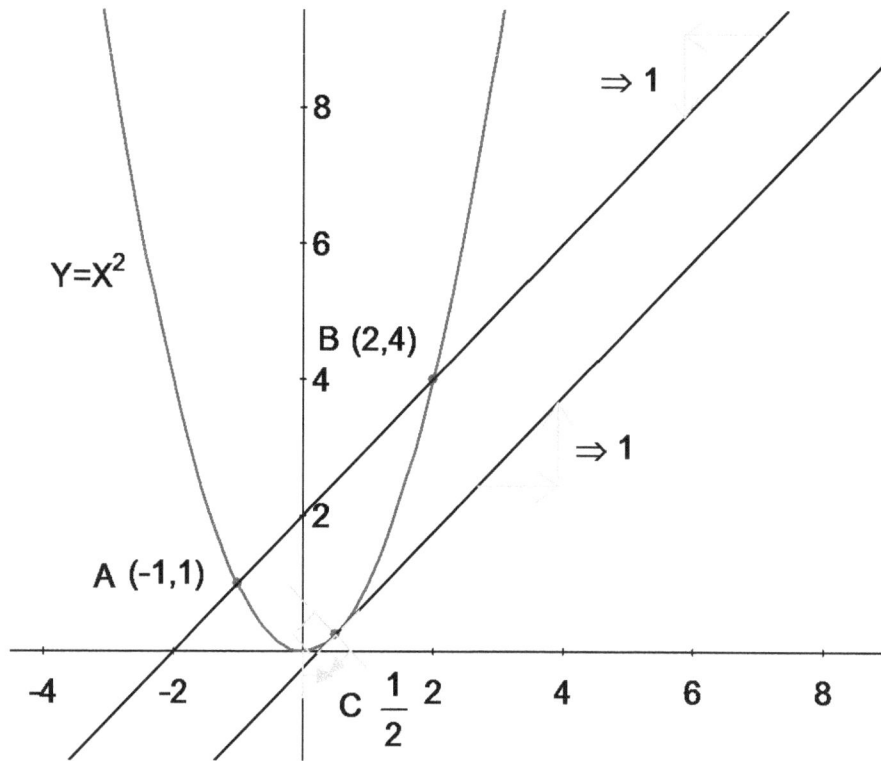

$Y=X^2$

$\Rightarrow 1$

8

6

B (2,4)

4

$\Rightarrow 1$

2

A (-1,1)

-4   -2   C $\dfrac{1}{2}$ 2   4   6   8

b. $y = \sqrt{x}$ on [1, 9]; $\dfrac{1}{2\sqrt{c}} = \dfrac{3-1}{9-1} = \dfrac{1}{4}$, so $c = 4$

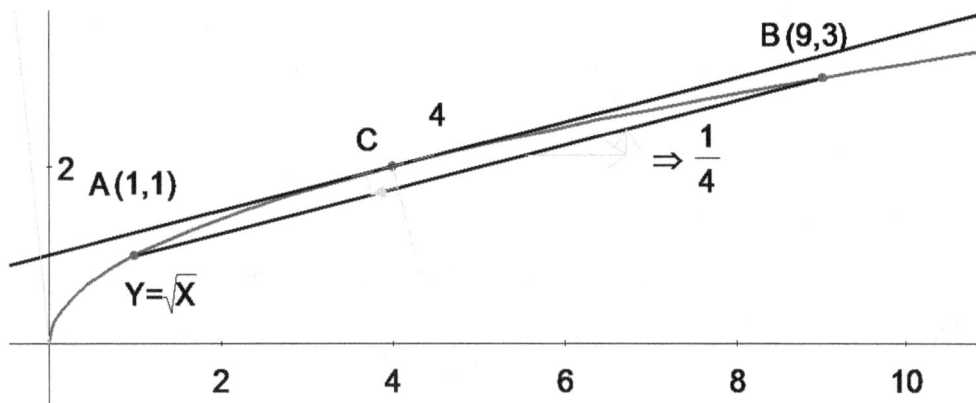

c. $y = x - 2\sin x$ on [-π, π], $1 - 2\cos(c) = \dfrac{\pi - (-\pi)}{2\pi} = 1$, so $c = -\dfrac{\pi}{2}, \dfrac{\pi}{2}$

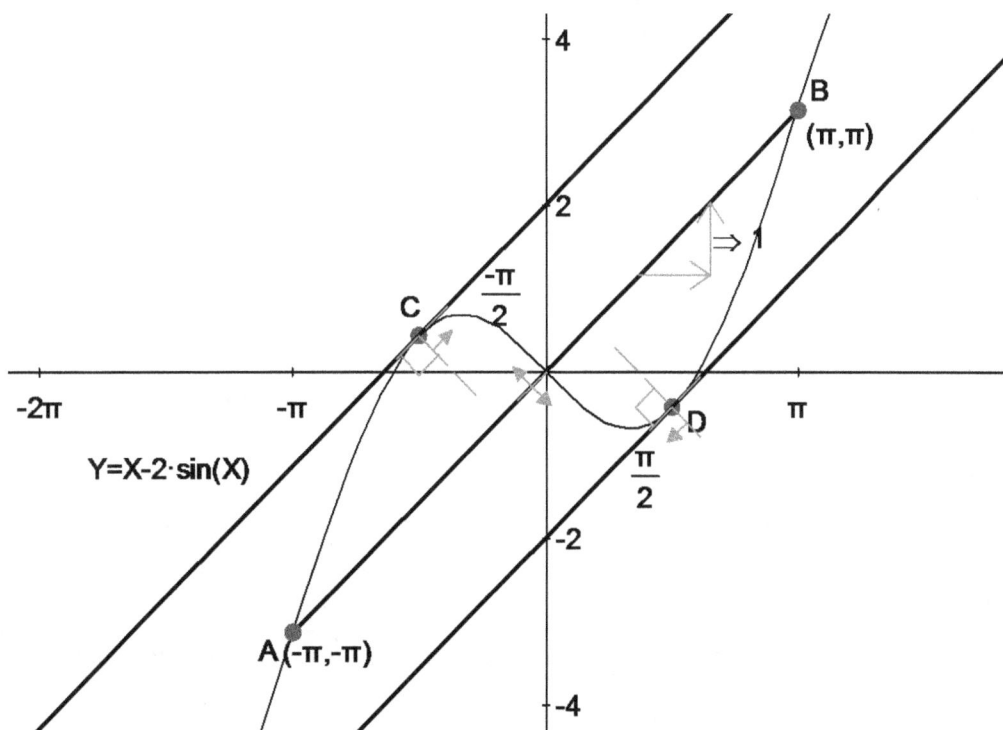

Mean Value Theorem

Exploration 2.2: explore and explain the Mean Value Theorem. If *f*(*x*) is defined and continuous on the interval [*a*, *b*] and differentiable on (*a*, *b*), then there is at least one number *c* in the interval (*a*, *b*) (that is *a* < *c* < *b*) such that $f'(c) = \dfrac{f(b) - f(a)}{b - a}$.

GENERAL STATEMENT

1. Draw a function *y* = *f*(*x*). Modify the shape of the function as needed.

2. Draw chord AB which will attach to the curve of the function when you constrain the coordinates of the endpoints as A(*a*, *f*(*a*)) and B(*b*, *f*(*b*)), so that *a* < *b*. Note: don't attach the endpoints to the curve when you draw the chord or your endpoints will be overconstrained.

3. Draw a tangent line to the graph of the function at an arbitrary point C (*c*, *f*(*c*)), so that *a* < *c* < *b*.

Q1. What is the slope of the chord AB?

4. Verify your answer using *Gx*.

Q2. What is the slope of the tangent line through point C?

5. Verify your answer using *Gx*.

Q3. Is it possible to find the location of point C on the graph of *f*(*x*) between points A and B so that the slope of the tangent line is equal to the slope of the chord AB?

Q4. Is it possible to find more than one location of the point C that satisfies this condition? (Reminder: [*a*, *b*] is an arbitrary interval such that *f*(*x*) is continuous on the [*a*, *b*] and differentiable on (*a*, *b*))

Q5. What would be the relative position of the chord to the tangent line(s) in this case?

APPLICATIONS

Q6. For the functions and intervals provided below use analytical and graphical methods to

1) Graph the given function on the given interval.

2) Find and graph the secant line through the endpoints of the given interval.

3) Find all points on the given interval where the function satisfies the Mean Value Theorem and graph the corresponding tangent lines.

a. $y = x^2$ on $[-1, 2]$; b. $y = \sqrt{x}$ on $[1, 9]$; c. $y = x - 2\sin x$ on $[-\pi, \pi]$.

2.3 The Derivative of Even Functions

<u>Exploration 2.3:</u> Given a differentiable function, *f(x)*, such that the graph of *f(x)* is symmetrical relative to the *y*-axis, i.e. *f(x)* is an even function. Does the graph of *f'(x)* have symmetry?

SUMMARY

<u>Mathematics Objectives:</u>

- Explore the symmetry of a derivative of an even function.

- Justify that the derivative of an even function is an odd function using basic derivative rules.

<u>Vocabulary:</u>

- Reflection symmetry

- Central (rotational by 180°) symmetry

- Signum (or sign) function

<u>Pre-requisites:</u>

- Symmetry of graphs of even and odd functions

- Basic derivative rules

- Derivative as a slope of a tangent line

- $\text{sign}(x) = \begin{cases} \dfrac{x}{|x|}, x \neq 0 \\ 1, x = 0 \end{cases}$

<u>Problem Notes:</u>

- In this exploration students explore the symmetry of the graph of a derivative of an even function. Students first explore the special case of a function $y = x^{2n}$. They find the derivative of the function by using both algebraic (differentiation) and geometric (slope of the tangent line) methods. Then they explore the symmetry, again using algebraic (differentiation) and geometric (transformation) methods.

- Students explore the symmetry of the derivative of a generic even function $y = f(|x|)$ applying the same methods as above.

- This exploration provides an opportunity for students to work with abstract concepts of functions and their derivatives. Graphing $y = f(|x|)$ and symbolically creating a graph of the derivative by using slopes of tangent lines to the graph of $y = f(|x|)$ is a powerful feature of *Gx*.

Technology skills:

- Draw: function
- Constrain: coordinates of a point, a point proportional along a curve
- Construct: tangent, rotation
- Calculate: slope of a line

Extension:

Given an even differentiable function $y = f(x)$, which of the following can be true:

a. $f'(0) < 0$, b. $f'(0) = 0$, c. $f'(0) > 0$?

STEP-BY-STEP INSTRUCTIONS

SPECIAL CASE $y = x^{2n}$

1. Draw a function $y = x^{2n}$ (*n* is an integer).

   a. Use **Toggle grid and axes** to show the axes without a grid.

   b. Choose **Draw** → Function. Select Type → Cartesian. In the Y = prompt type x^(2*n).

   c. Check the value of *n* in the **Variables** toolbox and make sure that $n \geq 2$.

Q1. Use algebra to explain why this function is even.

A1: $f(-x) = (-x)^{2n} = (-1)^{2n} x^{2n} = x^{2n} = f(x)$

2. Draw a tangent line to the graph of the function at an arbitrary point A.

   a. Select the plot of the function and choose **Construct** → Tangent. The tangent line and the point of tangency (point A) will be displayed.

b. Select point A and the graph of *f*(*x*) and choose **Constrain** → Point proportional along curve. Type *x* in the edit box to constrain the x-coordinate of point A to be *x*.

3. Calculate the slope of the tangent line to *f(x)* at the point A.

a. Select the tangent line and choose **Calculate** → Slope.

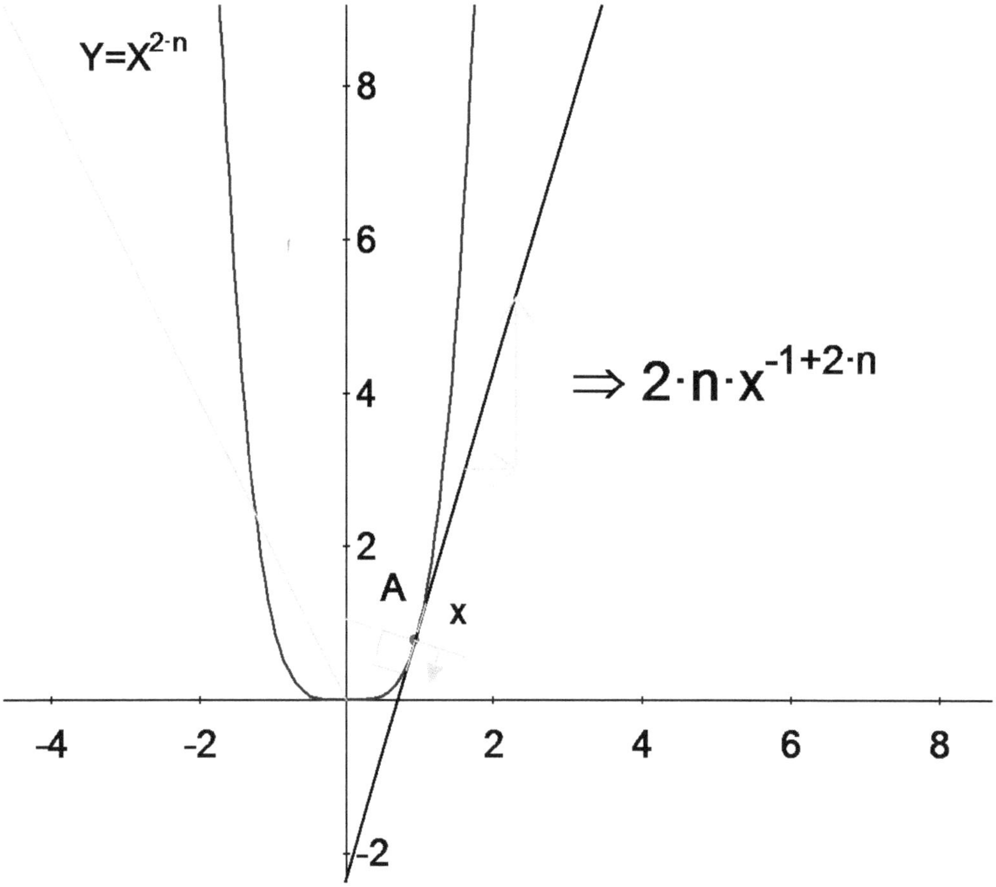

Y=X^{2·n}
8
6
4
2
A
x
⇒ 2·n·x^{-1+2·n}
-4
-2
2
4
6
8
-2

Q2. Use calculus to explain why this expression for the slope of the tangent line is correct.

A2: $f'(x) = \left(x^{2n}\right)' = 2nx^{2n-1}$

4. Graph the slope of the tangent line.

a. Click on the expression for the slope of the tangent line and choose **Edit** → Copy As → String.

b. Open a new document.

c. Use **Toggle grid and axes** to show the axes without a grid.

d. Choose **Draw** → Function.

e. In the open edit box select Cartesian for Type. Clear the Y = prompt (or just select it) and paste the expression from the clipboard (Ctrl – V). Press OK.

f. Double check your value of n in the **Variables** toolbox and make sure that $n \geq 2$.

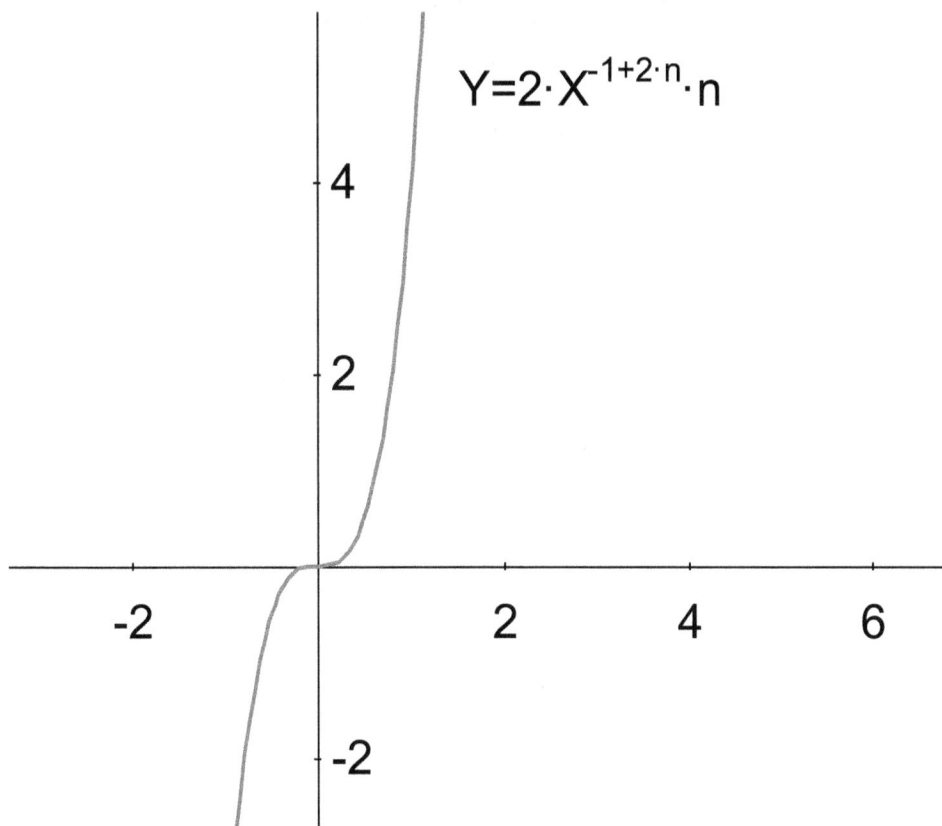

$$Y = 2 \cdot X^{-1+2 \cdot n} \cdot n$$

Q3. Does this graph have symmetry? Explain using two methods: a) analytical and b) graphical

A3: *The graph of the function has central symmetry about the origin. Analytically* $f'(-x) = 2n(-x)^{2n-1} = 2n(-1)^{2n-1}x^{2n-1} = -2nx^{2n-1} = -f'(x)$, *so the derivative is an odd function. Graphically, use a rotation by 180° around the origin and confirm the central symmetry.*

a. Click the graph of the function and choose **Construct** → Rotation. Make the center of the rotation the origin.

b. Type π in the open window for the angle of rotation (if the angle is set in degrees, type 180°). Press enter.

c. The graph of the image will coincide with the original graph

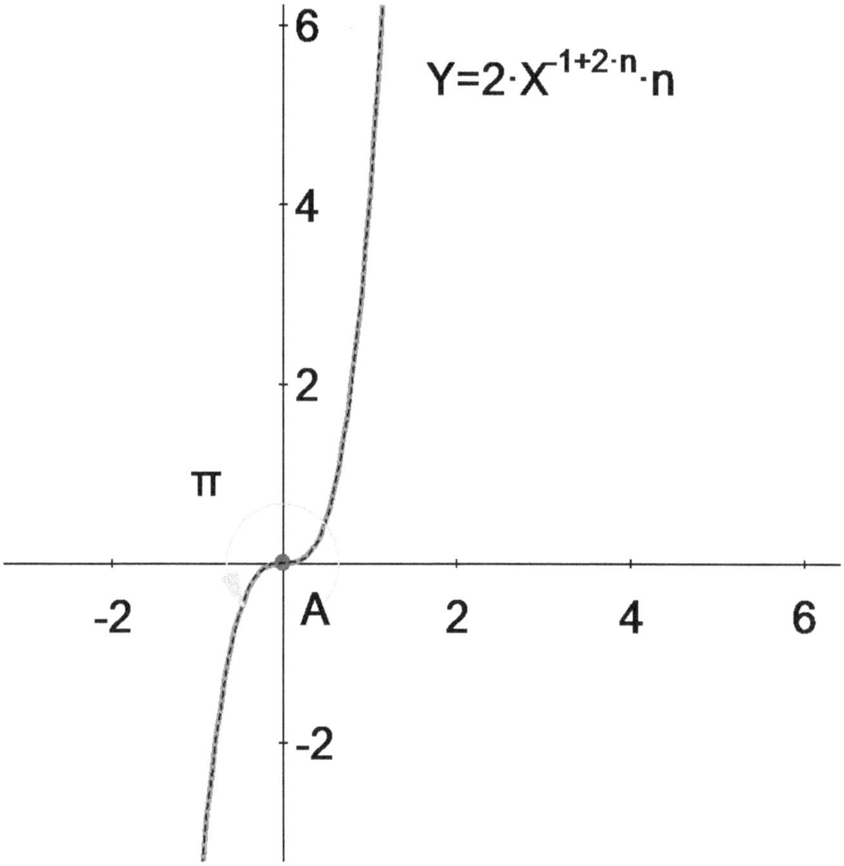
6
Y=2·X^1+2·n·n
4
2
π
-2
A
2
4
6
-2

GENERAL CASE $y = f(|x|)$

5. In a new document graph a function $y = f(|x|)$.

a. Use **Toggle grid and axes** to show the axes without grid.

b. Choose **Draw** → Function. Select Type → Cartesian. In the Y = prompt type f(abs(x)).

Q4. Explain why this function is even

A4: $f(|-x|) = f(|x|)$, *since* $|-x| = |x|$.

Q5. Will the graph of $f'(|x|)$ have symmetry? Explain.

A5: *Since this function is an even function, the derivative of this function should be odd.*

$$\frac{df(|x|)}{dx} = f'(|x|) \cdot \frac{d|x|}{dx} = \text{signum}(x) \cdot f'(|x|), \text{ where signum}(x) = \frac{x}{|x|}.$$

6. Verify your results:

   a. Check **Edit → Preferences → Math → Output →** Use Assumptions should be set to False.

   b. Draw a tangent line to the graph of $f(|x|)$ at a point A.

      • Select the plot of the function and choose **Construct →** Tangent.

      • Select point A and graph of $f(|x|)$ and choose **Constrain →** Point proportional along curve. Type x in the edit box to constrain x-coordinate of the point A to be x.

   c. Calculate the slope of the tangent line at point A.

      • Select the tangent line and choose **Calculate →** Slope

   d. Graph the slope of the tangent line.

      • Click on the expression for the slope of the tangent line and choose **Edit →** Copy As → String.

      • In the open edit box select Cartesian for Type. Clear the Y = prompt and paste the expression from the clipboard (Ctrl – V). Press OK.

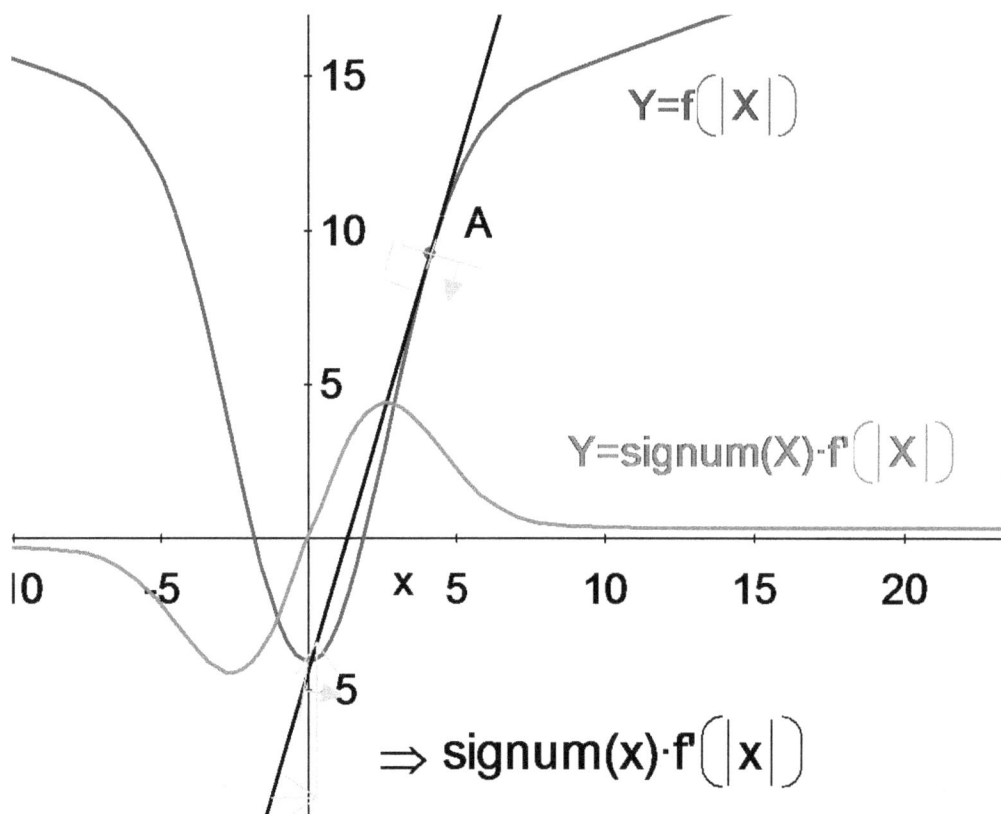

$$Y=f\left(|X|\right)$$

A

$$Y=signum(X)\cdot f'\left(|X|\right)$$

$$\Rightarrow signum(x)\cdot f'\left(|x|\right)$$

Q6. Given $f(-x) = f(x)$. Justify the symmetry of the graph of $f'(x)$.

A6: *The graph of the function has rotational symmetry about the origin. Analytically*

$$\frac{d}{dx}f(-x) = \frac{d}{dx}f(x) \Leftrightarrow -f'(-x) = f'(x) \Leftrightarrow f'(-x) = -f'(x),$$ *so the derivative is an odd function.*

Graphically, use a rotation by 180° around the origin and confirm the rotational symmetry.

a. Click on the graph of the function and choose **Construct** → Rotation. Make the center of the rotation the origin.

b. Type π in the open window for the angle of rotation.

c. The graph of the image will coincide with the original graph

The Derivative of Even Functions

<u>Exploration 2.3:</u> Given a differentiable function, $f(x)$, such that the graph of $f(x)$ is symmetrical relative to the y-axis, i.e. $f(x)$ is an even function. Does the graph of $f'(x)$ have symmetry?

SPECIAL CASE $y = x^{2n}$

1. Draw a function $y = x^{2n}$ (n is integer).

Q1. Use algebra to explain why this function is even.

2. Draw a tangent line to the graph of the function at an arbitrary point A (x, y).

3. Calculate the slope of the tangent line to $f(x)$ at the point A.

Q2. Use calculus to explain why this expression for the slope of the tangent line is correct.

4. Graph the slope of the tangent line.

Q3. Does this graph have symmetry? Explain using two methods: a) analytical and b) graphical

GENERAL CASE $y = f(|x|)$

5. In a new document graph a function $y = f(|x|)$.

Q4. Explain why this function is even

Q5. Will the graph of $f'(|x|)$ have symmetry? Explain.

6. Verify your results:

    a. Check **Edit → Preferences → Math → Output →** Use Assumptions should be set to False.

    b. Draw a tangent line to the graph of $f(x)$ at a point A (x, y).

    c. Calculate slope of the tangent line at point A.

    d. Graph slope of the tangent line.

Q6. Given $f(-x) = f(x)$. Justify symmetry of graph of $f'(x)$.

2.4 The Derivative of Odd Functions

Exploration 2.4: Given a differentiable function $g(x)$, such that the graph of $g(x)$ has a central symmetry relative to the origin of the system of coordinates, *i.e. g(x)* is an odd function. Does the graph of $g'(x)$ have symmetry?

SUMMARY

Mathematics Objectives:

- Explore the symmetry of a derivative of an odd function.

- Justify that the derivative of an even function is an even function using basic derivative rules.

Vocabulary:

- Reflection symmetry

- Rotational symmetry

Pre-requisites:

- Symmetry of graphs of even and odd functions

- Basic derivative rules

- Derivative as a slope of a tangent line

- $\text{sign}(x) = \begin{cases} \dfrac{x}{|x|}, x \neq 0 \\ 1, x = 0 \end{cases}$

Problem Notes:

- In this exploration students explore the symmetry of the graph of a derivative of an odd function. Students first explore the special case of a function $y = x^{2n+1}$. They find the derivative of the function by using both, algebraic (differentiation) and geometric (slope of the tangent line) methods. Then they explore the symmetry, again using algebraic (differentiation) and geometric (transformation) methods.

- Then the students explore symmetry of the derivative of a generic odd function $y = \text{signum}(x)g(|x|)$ applying the same methods as above.

- This exploration provides an opportunity for students to work with abstract concepts of functions and their derivatives. Graphing $y = \text{signum}(x)g(|x|)$ and symbolically creating a graph of the derivative by using slopes of tangent lines to the graph of $y = \text{signum}(x)g(|x|)$ is a powerful feature of the *Geometry Expressions*.

Technology skills:

- Draw: function, tangent line to curve
- Constrain: coordinates of a point, a point proportional along a curve
- Calculate: slope of a line
- Construct: rotation

Extension:

Prove that the derivative of a periodic function is periodic. Does the converse statement hold true? (i. e. Prove or disprove that the derivative of a non-periodic function is non-periodic.)

STEP-BY-STEP INSTRUCTIONS

SPECIAL CASE $y = x^{2n+1}$

1. Draw an odd function $g(x) = x^{2n+1}$ (n is integer).

    a. Use **Toggle grid and axes** to show the axes without a grid.

    b. Choose **Draw** → Function. Select Type → Cartesian. In the Y = prompt type x^(2*n+1).

    c. Check the value of n in the Variables view and make sure that $n \geq 2$.

Q1. Use algebra to explain why this function is odd.

A1: $g(-x) = (-x)^{2n+1} = (-1)^{2n+1}x^{2n+1} = -x^{2n+1} = -g(x)$

2. Draw a tangent line to the graph of *g(x)* at a point A.

    a. Select the plot of the function and choose **Construct** → Tangent. The tangent line and point of tangency (point A) will be displayed.

    b.   Select point A and the graph of $g(x)$ and choose **Constrain** → Point proportional along curve. Type x in the edit box to constrain the x-coordinate of point A to be x.

3.   Calculate the slope of the tangent line at point A.

    a.   Select the tangent line and choose **Calculate** → Slope.

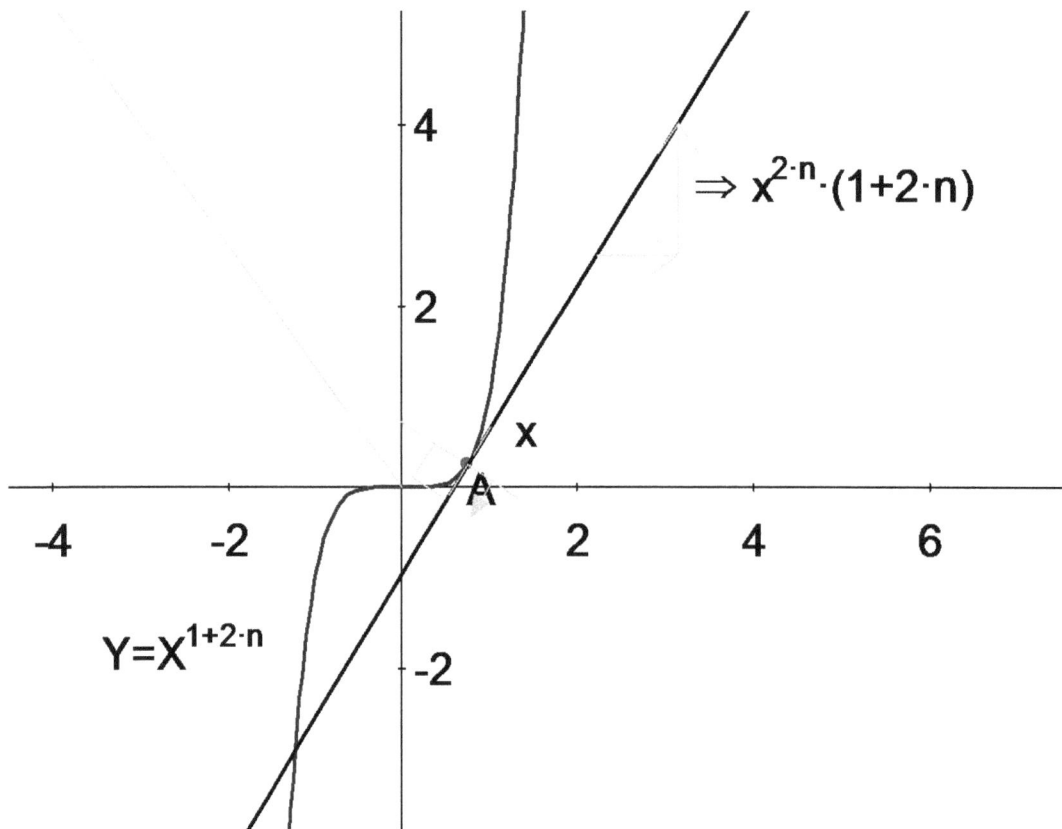

Q2. Use calculus to explain why this expression for the slope of the tangent line is correct.

A2: $g'(x) = \left(x^{2n+1}\right)' = (2n+1)x^{2n}$

4.   Graph the slope of the tangent line.

    a.   Click on the expression for the slope of the tangent line and choose **Edit** (or right click) → Copy As → String.

    b.   Open a new document.

c. Use **Toggle grid and axes** to show the axes without a grid.

d. Choose **Draw** → Function.

e. In the open edit box select Cartesian for Type. Clear the Y = prompt and paste the expression from the clipboard (Ctrl – V). Press OK.

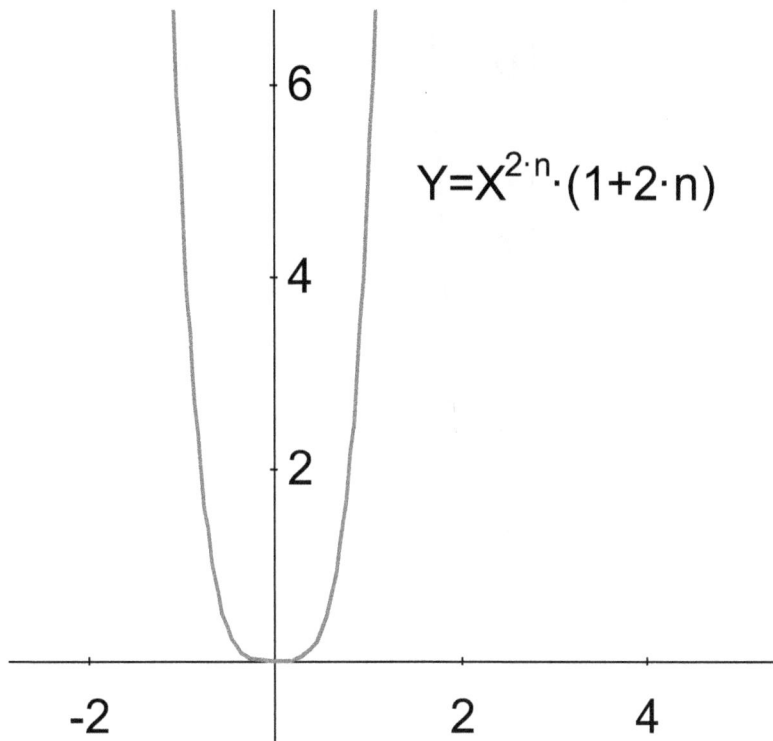

$$Y=X^{2 \cdot n} \cdot (1+2 \cdot n)$$

Q3. Does this graph have symmetry? Explain using two methods: a) analytical and b) graphical

A3: *The graph of the function has reflection symmetry over the y-axis. Analytically,*
$$g'(-x)=(2n+1)(-x)^{2n}=(2n+1)(-1)^{2n}x^{2n}=(2n+1)x^{2n}=g'(x),$$ *so the derivative is an even function. Graphically, use reflection about y-axis and confirm reflective symmetry.*

a. Click on the graph of the function and choose **Construct** → Reflection. Click on the y – axis.

b. The graph of the image coincides with the original graph.

$$Y=X^{2 \cdot n} \cdot (1+2 \cdot n)$$

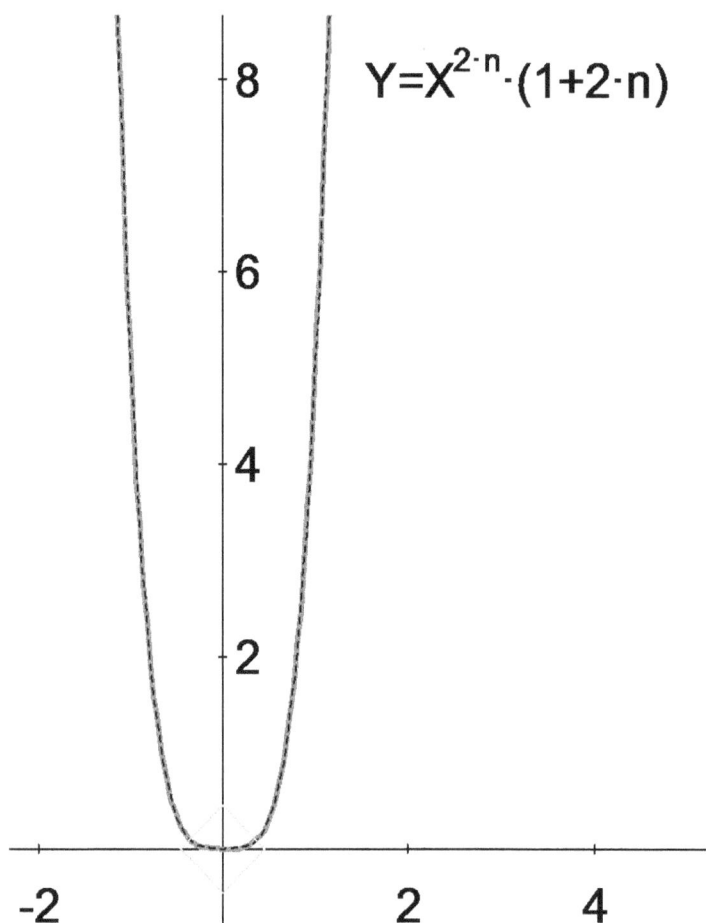

GENERAL CASE $y = \text{signum}(x) \cdot g(|x|)$

5. In a new document graph an odd function $y = \text{signum}(x) \cdot g(|x|)$.

   c. Use **Toggle grid and axes** to show the axes without grid.

   d. Choose **Draw** $\rightarrow$ Function. Select Type $\rightarrow$ Cartesian. In the Y = prompt type signum(x)*g(abs(x)).

Q4. Why is this function odd? Explain.

A4: $\text{signum}(-x) \cdot g(|-x|) = -\text{signum}(x) \cdot g(|x|),$ *since* $|-x| = |x|$ *and* $\text{signum}(-x) = -\text{signum}(x).$

Q5. Will the graph of the derivative of this function have symmetry? Explain

A5: Since this function is an odd function, as we determined above, the derivative of this function should be even.

$$\frac{d}{dx}\left[\text{signum}(x)\cdot g(|x|)\right] = \text{signum}(x)\cdot \frac{d}{dx} g(|x|) = \text{signum}(x)\cdot g'(|x|)\cdot \frac{d|x|}{dx} = (\text{signum}(x))^2 \cdot g'(|x|)$$

where $\text{signum}(x) = \dfrac{x}{|x|}.$

6. Verify your results:

   a. Draw a tangent line to the graph of this function at a point A (x, y).

   - Select the plot of the function and choose **Construct** → Tangent to Curve.

   - Select point A and graph of *g(x)* and choose **Constrain** → Proportional. Type x in the edit box to constrain the x-coordinate of the point A to be x.

   b. Calculate the slope of the tangent line at point A.

   - Select the tangent line and choose **Calculate** → Slope

   c. Graph the slope of the tangent line.

   - Click on the expression for the slope of the tangent line and choose **Edit** → Copy As → String.

   - In the open edit box select Cartesian for Type. Clear the Y = prompt and paste the expression from the clipboard (Ctrl – V). Press OK.

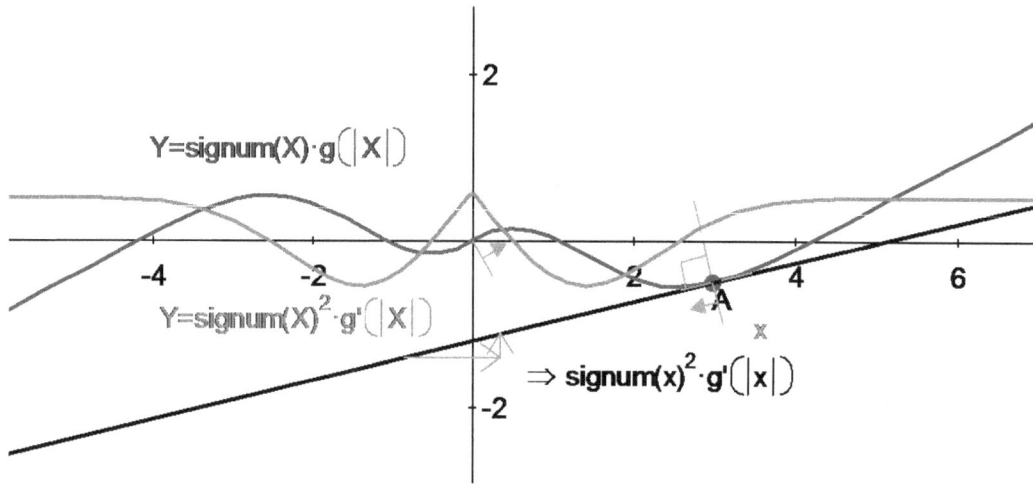

Q6. Given $g(-x) = -g(x)$. Justify the symmetry of the graph of $g'(x)$.

A6: Differentiate both sides of the equation:

$$\frac{d}{dx}g(-x) = \frac{d}{dx}(-g(x)) \Leftrightarrow -g'(-x) = -g'(x) \Leftrightarrow g'(-x) = g'(x)$$

Derivative of Odd Functions

<u>Exploration 2.4:</u> Given a differentiable function $g(x)$, such that the graph of $g(x)$ is symmetrical relative to the origin of the system of coordinates, *i.e.* $g(x)$ is an odd function. Does the graph of $g'(x)$ have symmetry?

SPECIAL CASE $y = x^{2n+1}$

1.   Draw an odd function $g(x) = x^{2n+1}$ (n is integer).

Q1. Use algebra to explain why this function is odd.

2.   Draw a tangent line to the graph of $g(x)$ at a point A.

3.   Calculate the slope of the tangent line to the graph of $g(x)$ at point A.

Q2. Use calculus to explain why this expression for the slope of the tangent line is correct.

4.   Graph the slope of the tangent line.

Q3. Does this graph have symmetry? Explain using two methods: a) analytical and b) graphical.

GENERAL CASE $y = \text{signum}(x) \cdot g(|x|)$

5.   In a new document graph an odd function $y = \text{signum}(x) \cdot g(|x|)$.

Q4. Why is this function odd? Explain.

Q5. Will the graph of the derivative of this function have symmetry? Explain

6.   Verify your results:

   a.   Draw a tangent line to the graph of this function at a point A.

   b.   Calculate the slope of the tangent line at point A.

   c.   Graph the slope of the tangent line.

Q6. Given $g(-x) = -g(x)$. Justify symmetry of graph of $g'(x)$.

2.5 Differentiability of a Piecewise Function at a Point

<u>Exploration 2.5:</u> Given the function: $h(x) = \begin{cases} f(x), x < 0 \\ g(x), x > 0 \\ C, x = 0 \end{cases}$, where C = const. What are the

necessary and sufficient conditions for the function $h(x)$ to be differentiable at $x = 0$?

SUMMARY

<u>Mathematics Objectives:</u>

- Explore graphically the differentiability of a piecewise function at a joint point.

- Determine that in order for a piecewise function to be differentiable at a joint point, both the function and its derivative must be continuous at a joint point.

<u>Vocabulary:</u>

- Continuity

- Differentiability

- One-side limit

- Piecewise function

- Joint point

<u>Pre-requisites:</u>

- Graph of a piecewise function.

- Continuity of a function.

- Limit of a function at a point.

- Derivative as a slope of a tangent line.

<u>Problem Notes:</u>

- Students frequently confuse differentiability with continuity. For example, $y = |x|$ is continuous at 0 but is not differentiable at 0. Why is this true? This exploration provides students with the opportunity to explore the continuity and differentiability visually and symbolically.

- Students first explore a specific piecewise function. They graph the function and manipulate the graph of the function to determine that in order for the function to be continuous at a joint point, it has to be defined at the joint point, the left-side and the right-side limits of the function have to be equal to each other at a joint point, and this limit has to be equal to the value of the function at this point.

- Students then explore tangent lines to the graph of the function as the points of tangency approach the joint point from the left and from the right. Students determine that the slopes of tangent lines on the left and on the right have to be equal in order for the continuous function to be also differentiable. Thus, they establish the fact that the derivative of the function must be also continuous. In conclusion students generalize their findings for the case of a generic piecewise function.

- As with any graphing utility, you have to be careful of domain restrictions. *Geometry Expressions* can graph a function only on a closed interval; thus, both parts of the piecewise function $h(x)$ are graphed on intervals that include 0. However, students should understand that even though the software can provide a value of $x+1$ or e^{kx} at $x=0$, these are not defined at $x=0$. For this reason we defined the piecewise function $h(0)$ separately at a joint point. This should also help students clarify their understanding of continuity as not only the existence of left-side and right-side limits at the joint point, but also that these limits should be equal to the value of the function at the same point.

- One of the powerful attributes of *Geometry Expressions* is that students can move points, drag functions and observe the behavior of tangent lines on the left side and the right side of the joint point so that the tangent lines coincide, hence, the function is differentiable.

Technology skills:

- Draw: function, point
- Construct: tangent line
- Constrain: point proportional along the curve
- Calculate: slope

Extension:

Given the piecewise function $y = \begin{cases} 0, x \leq 0 \\ f(x), 0 < x < 1 \\ x, x \geq 1 \end{cases}$. Find a function $f(x)$ defined on the interval (0, 1) so that the piecewise function is differentiable.

STEP-BY-STEP INSTRUCTIONS

DIFFERENTIABILITY OF $h(x) = \begin{cases} x+1, x < 0 \\ e^{kx}, x > 0 \\ 1, x = 0 \end{cases}$ AT X=0

1. Draw a graph of the function by separately plotting each part of the function on its sub-domain.

    a. Use **Toggle grid and axes** to show the axes without a grid.

    b. Choose **Draw** → Function. Select Type → Cartesian. In the Y = prompt type x+1. Select Start Value -10 and End Value 0. Press OK.

    c. Choose **Draw** → Function. Select Type → Cartesian. In the Y = prompt type exp(k*x). Select Start Value 0 and End Value 10. Press OK.

    d. Choose **Draw** → Point and plot point C on the y – axis. Select the point C and y-axis and choose **Constrain** → Proportional along the curve. In the open edit box type 1.

Note: as with any graphical utility, Geometry Expressions can graph functions only on a closed interval. Thus, both parts of the function h(x) are graphed on intervals that include 0. However, students should understand that even though the software can provide the value of $x+1$ or e^{kx} at x = 0, these are not defined at x = 0. This is the reason we defined the piecewise function h(0) separately as a point. It will also help students to clarify their understanding of continuity as not only the existence of left-side and right-side limits at the point, but also the fact that these limits should be equal to the value of the function.

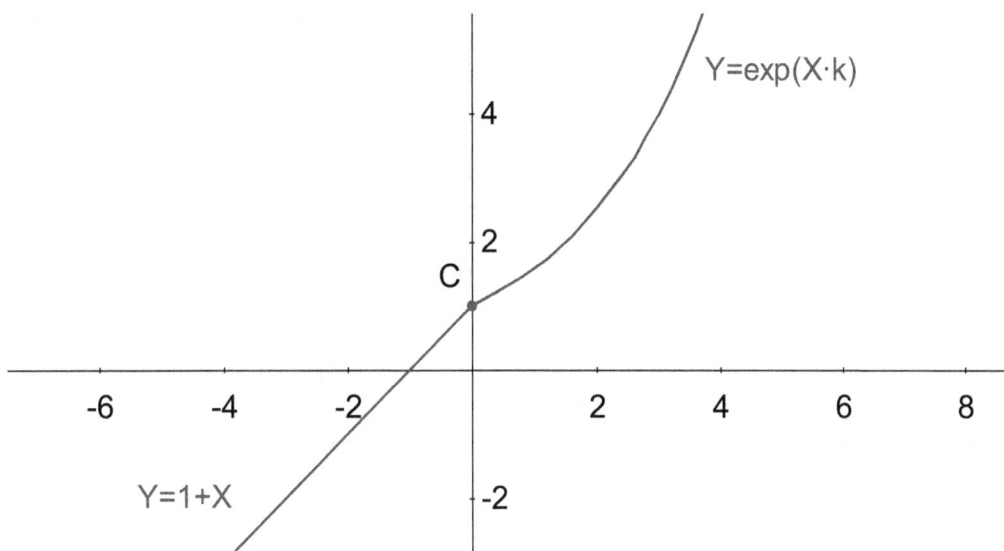

2. Drag the exponential branch of the function and observe its behavior at $x = 0$.

Q1. Confirm that the function is continuous at $x = 0$.

A1: $\lim_{x \to 0^-}(x+1) = 1 = \lim_{x \to 0^+} e^{kx} = h(0)$.

Q2: What is the derivative of the function?

A2: $\dfrac{dy}{dx} = \begin{cases} 1, x < 0 \\ ke^{kx}, x > 0 \end{cases}$, it is not defined at $x = 0$ for all values of k.

3. Confirm your calculations with the help of the software.

   a. Select the exponential piece of the function and choose **Construct** → Tangent to Curve. The tangent line and the point of tangency will be displayed (point A).

   b. Select point A and the graph of the function and choose **Constrain** → Point proportional along curve. Type a in the open edit box.

   c. Select the tangent line and choose **Calculate** → Slope.

   d. Select the linear piece of the function and choose **Construct** → Tangent to Curve. The tangent line and the point of tangency will be displayed (point B).

   e. Select the tangent line and choose **Calculate** → Slope.

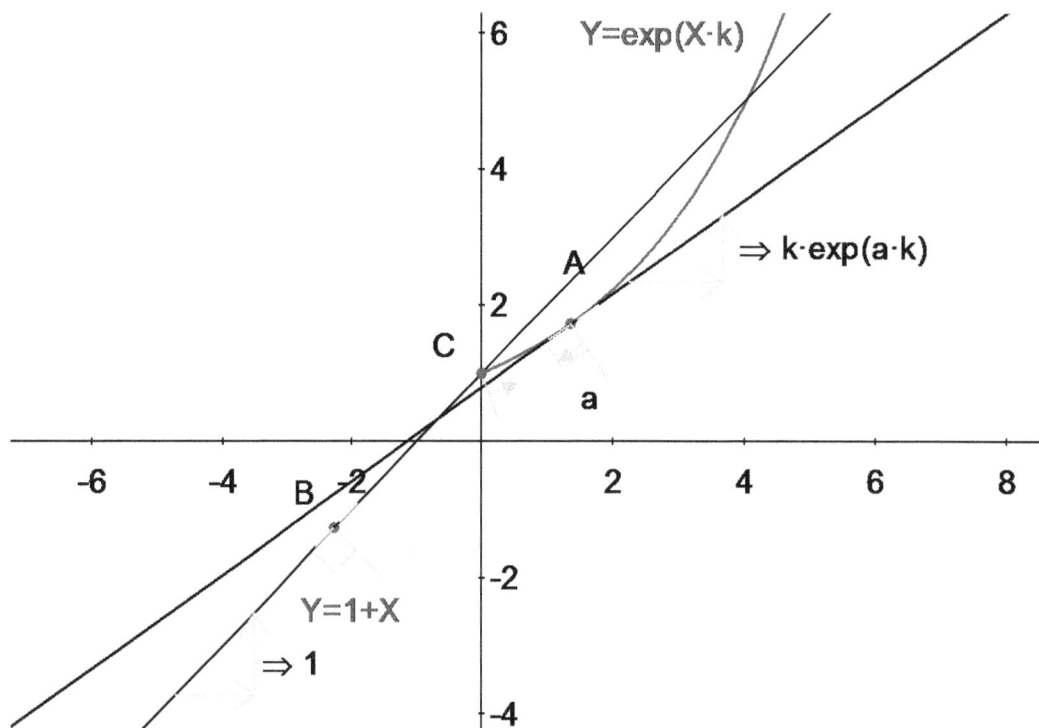

Q3. What is a necessary and sufficient condition on the tangent lines for the function to be differentiable at $x = 0$?

A3:   The slopes of the tangent lines of each piece should be equal at $x = 0$.

3.   Use the software to determine when this occurs.

*Note: students can move point A to point C and then drag the exponential curve or change the value of k using the slider in the **Variables** tool panel to find a value of k for which the tangent lines coincide.*

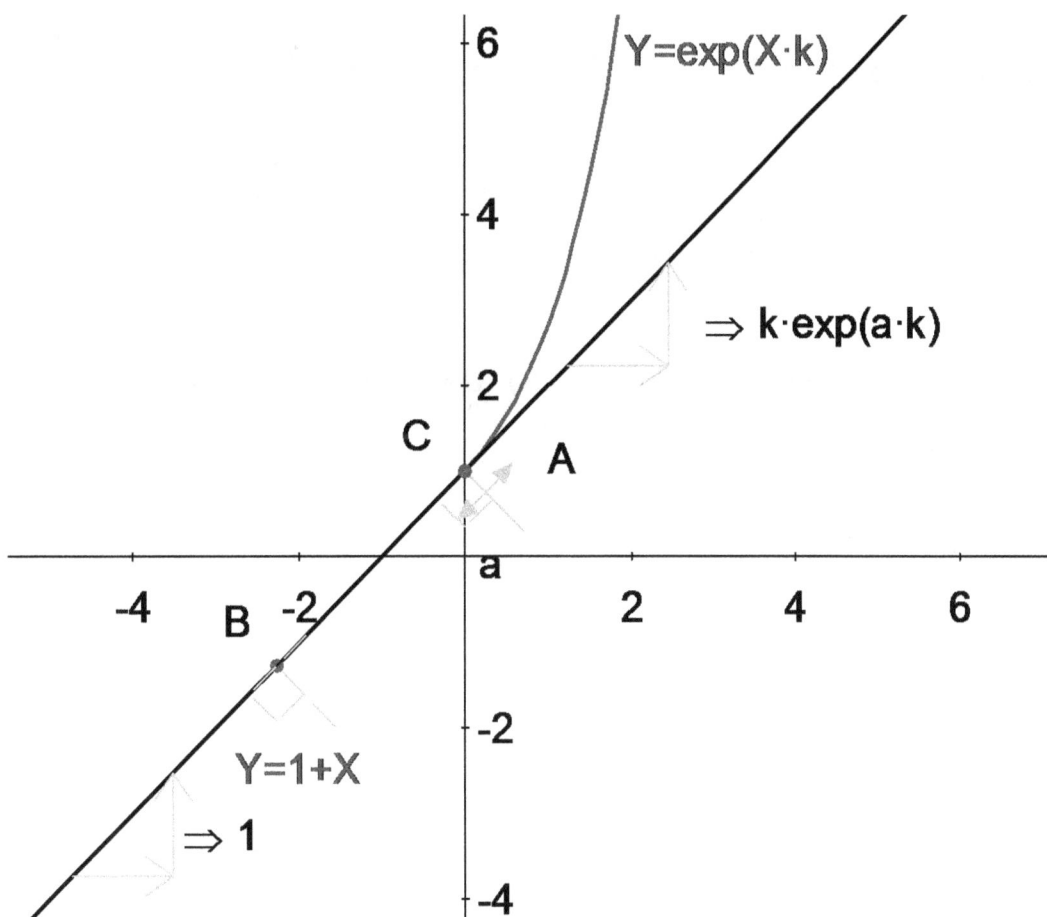

Q4. What is the value of k in this case?

A4: k = 1. Some students will find an approximate value, for example k = 0.998, so they will predict k =1 and then check it analytically.

Q5. Verify your answer algebraically.

A5: $\displaystyle\lim_{x\to 0^+} ke^{kx} = 1 \Rightarrow k = 1.$

DIFFERENTIABILITY OF $h(x) = \begin{cases} f(x), x < 0 \\ g(x), x > 0 \\ C, x = 0 \end{cases}$ AT X=0.

4. In the diagram you created using Gx, replace the specific functions with general functions such that $h(x) = \begin{cases} f(x), x < 0 \\ g(x), x > 0 \\ C, x = 0 \end{cases}$.

    a. Double-click on Y = x+1 and type f(x). Don't forget to make point B proportional along the curve.

    b. Double-click on Y = exp(k*x) and type g(x).

    c. Double-click on the *y*-coordinate (point proportional along the y axis) of point C and change it to c.

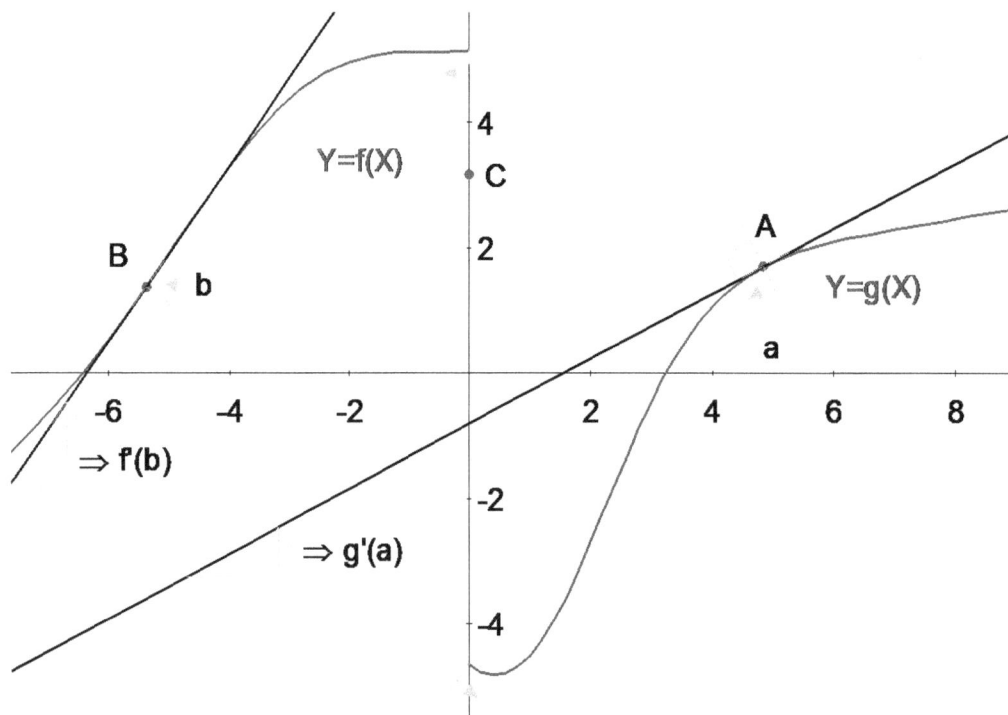

Q3. How can you make this function continuous at *x = 0*?

A3: Modify the curves so that $\lim_{x \to 0^-} f(x) = \lim_{x \to 0^+} g(x) = g(0)$. In other words, move the curves to connect at point C.

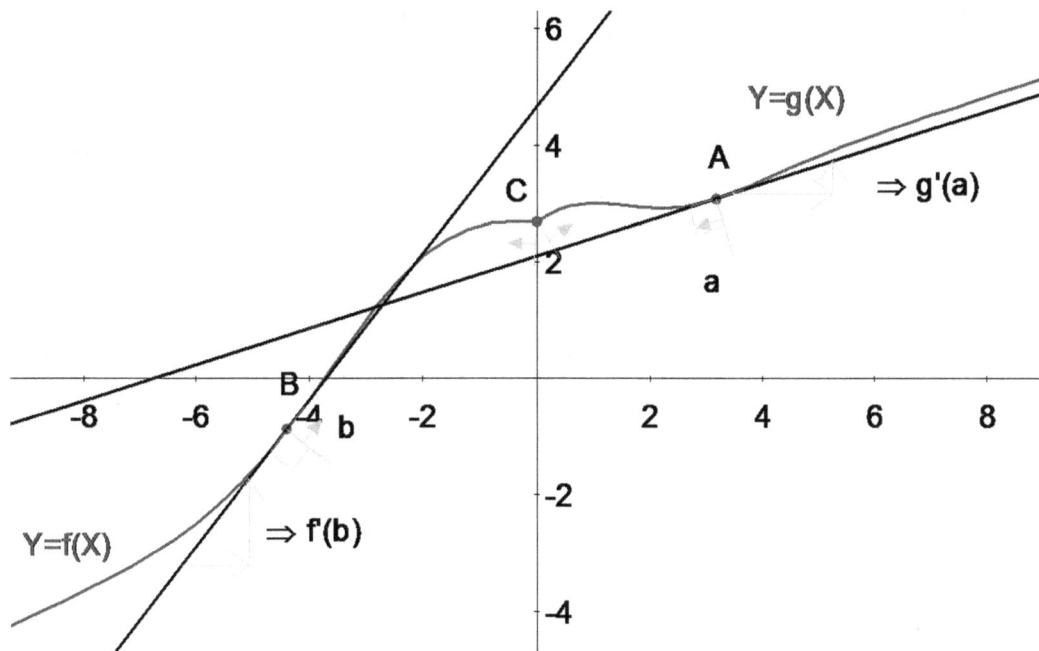

Note: In order to make the function h(x) continuous at x=0 with Gx, students must modify the shape of the function by following the steps below (this process may take a while, so the teacher may decide to skip the actual work with the software for the general case or develop a premade file in advance to give to the students):

a. Click the curve, hold the mouse and drag the curve just a little. A small circle, called a handle, with the name of the parameter will appear on the screen.

b. Similarly, click in another part of the function and drag it.

c. There are a total of 5 different parameters that are responsible for the shape of the function, and you can change each one of them by dragging each of the handles.

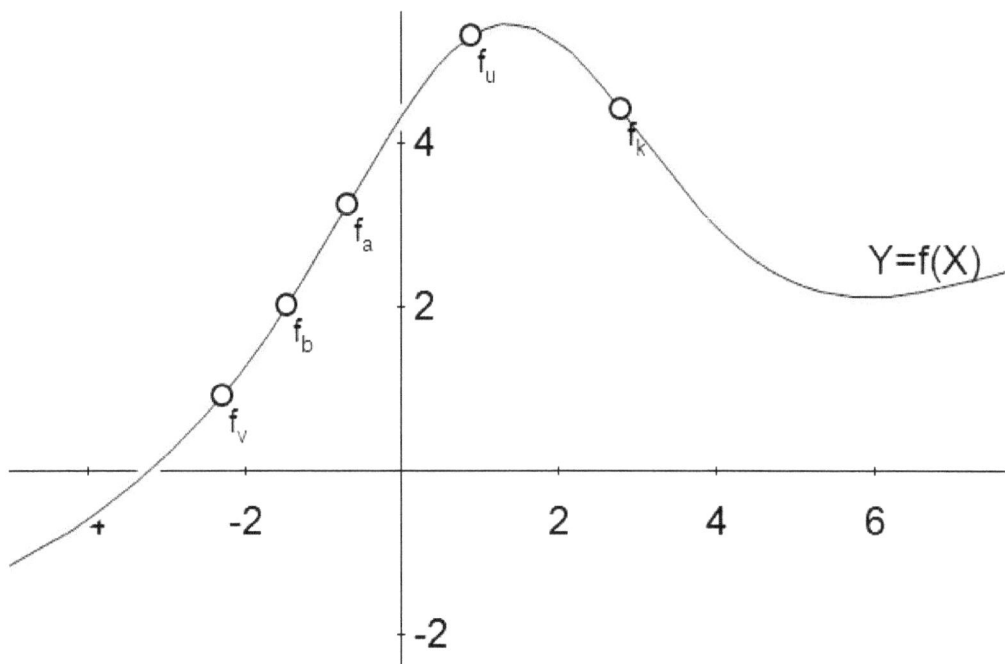

Q4. How can you make this function differentiable at $x = 0$?

A4: *Modify the curves so the slopes of the tangent lines at x = 0 are equal.*

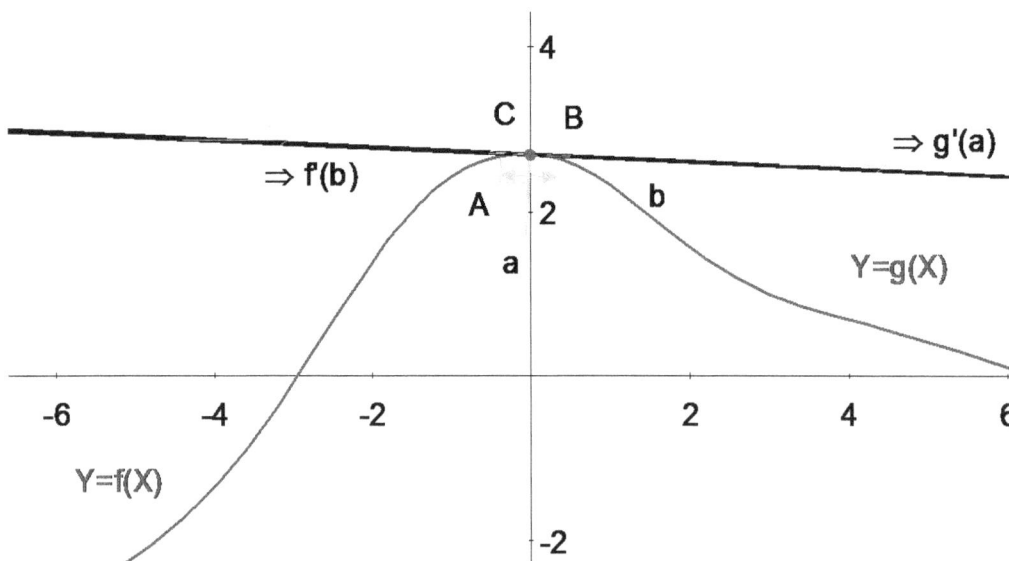

Q5. Can you formulate a general rule for the differentiability of piecewise functions at the joint point?

A5: A piecewise function is differentiable at the joint point if a) the function is continuous at the joint point, and b) the derivative from the right is equal the derivative from the left at the joint point.

Differentiability of a Piecewise Function at a Point

Problem Statement: Given the function: $h(x) = \begin{cases} f(x), x < 0 \\ g(x), x > 0 \\ C, x = 0 \end{cases}$, where C = const. What are the

necessary and sufficient conditions for the function $h(x)$ to be differentiable at $x = 0$?

DIFFERENTIABILITY OF $h(x) = \begin{cases} x+1, x < 0 \\ e^{kx}, x > 0 \\ 1, x = 0 \end{cases}$ **AT X=0**

1.   Draw a graph of the function by separately plotting each part of the function on its sub-domain.

2.   Drag the exponential branch of the function and observe its behavior at $x = 0$.

Q1.   Confirm that the function is continuous at $x = 0$.

Q2:   What is the derivative of the function?

3.   Confirm your calculations with the help of the software.

Q3.   What is a necessary and sufficient condition on the tangent lines for the function to be differentiable at $x = 0$?

3.   Use the Gx to determine when this occurs.

Q4.   What is the value of k in this case?

Q5.   Verify your answer algebraically.

DIFFERENTIABILITY OF $h(x) = \begin{cases} f(x), x < 0 \\ g(x), x > 0 \\ C, x = 0 \end{cases}$ **AT X=0**

4.   In the diagram you created using the software, replace the specific functions with

general functions such that $h(x) = \begin{cases} f(x), x < 0 \\ g(x), x > 0 \\ C, x = 0 \end{cases}$.

Q3. How can you make this function continuous at $x = 0$?

Q4. How can you make this function differentiable at $x = 0$?

Q5. Can you formulate a general rule for the differentiability of piecewise functions at the joint point?

2.6 Derivative of an Inverse Function

Exploration 2.6: Given a differentiable and monotonic function, *f(x)*. What is the relationship between the derivative of the function and its inverse?

SUMMARY

Mathematics Objectives:

- To discover that the derivative of a function and its inverse are reciprocals at the corresponding points, $\left(f^{-1}(y) \right)' = \dfrac{1}{f'(x)}$, where $y = f(x)$

- Use this relationship to determine the derivatives of specific functions.

Vocabulary:

- Monotonic function.

- Differentiable function.

- Inverse function.

Pre-requisites:

- Graphical and symbolic representations of inverse of a function.

- Existence of inverse functions.

- Reflection over the identity line $y = x$.

- Slope of a tangent line to a function as a derivative of a function.

Problem Notes:

- The relationship between the derivative of a function and its inverse function, when it exists, is important in determining the derivatives of certain functions. *Gx* allows students to clearly see the relationship between the slopes of tangents drawn on the function to its inverse and thus determine the formula for the derivative of the inverse of a function given the derivative of the function.

- In this activity students first explore the general case of an arbitrary monotonic function. The condition of monotonic is provided to assure the existence of an inverse function. Students graph an inverse function as a reflection of a graph of a given function over the identity line $y = x$, construct tangent lines to both graphs, and compare the slopes of these tangents. Thus, they establish general relationship between the derivative of a function and its inverse.

- Students then apply this relationship to find the derivative of the inverse of a specific function and confirm their calculations using the software.

- The reciprocal relationship between the derivative of a function and derivative of its inverse can be illustrated geometrically, as shown in the diagram below:

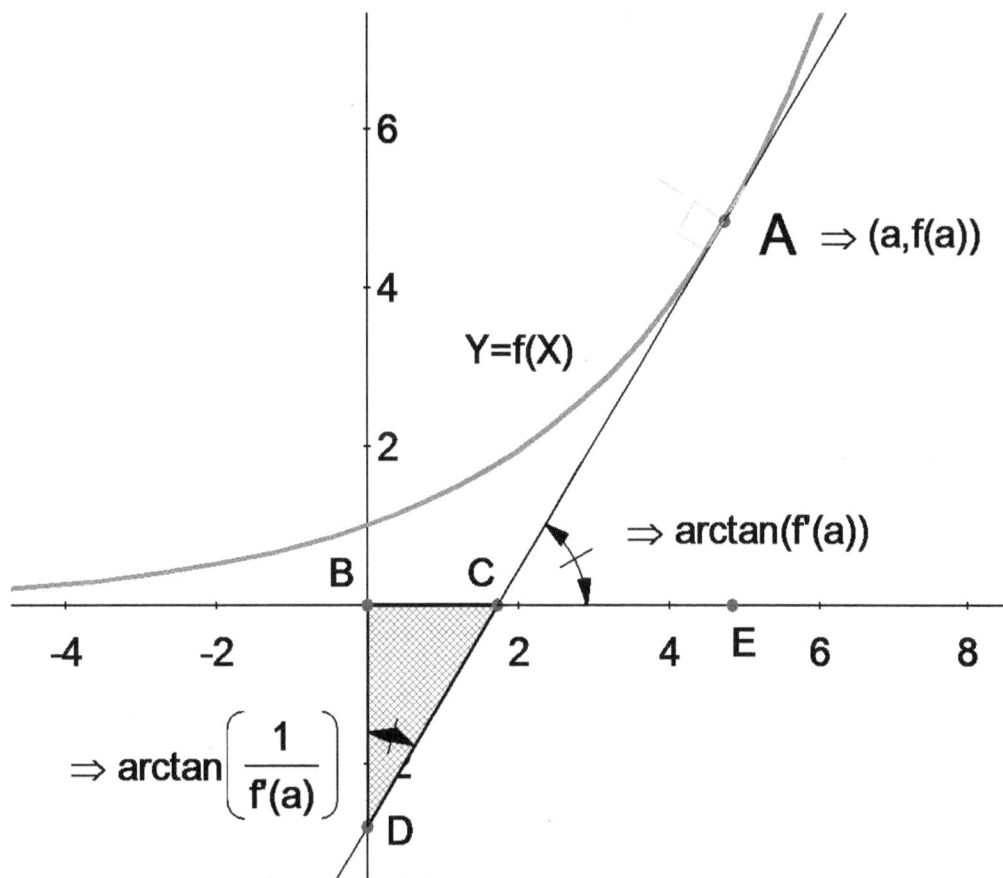

- Given the graph of the function $y = f(x)$, the derivative of the function at an arbitrary point A $(a, f(a))$ is equal to the slope of the tangent line to the graph of $f(x)$ at point A, which is the $\tan(\angle ACE)$. The inverse function equation can be written implicitly as $x = f(y)$. Thus, without changing variables, the graph of the inverse function is represented by the same curve, and has the same tangent line at the same point. However, the slope of the tangent line now is in relation to the positive direction of the *y*-axis, which means it is the tan(∠BDC). From ΔBDC it is clear that ∠DCB = 90°-∠BDC, and thus $\tan(\angle BDC) = \dfrac{1}{\tan(\angle DCB)}$. Since ∠DCB = ∠ACE and

$\tan(\angle DCB) = f'(x)\big|_{x=a}$ then $\left(f^{-1}(y)\right)'\big|_{y=f(a)} = \dfrac{1}{f'(x)\big|_{x=a}}$.

Technology skills:

- Draw: point, function

- Constrain: coordinates of a point, a point proportional along a curve, slope

- Calculate: slope, implicit equation

- Construct: reflection, tangent line to a curve

Extension:

Given the function $y = f(x)$. Find the derivative of $y = (f(x))^2$ at $x = a$.

STEP-BY-STEP INSTRUCTIONS

INVESTIGATION: THE GENERAL CASE

1. Draw a general function $y = f(x)$. Drag it at several places to make it either strictly increasing or strictly decreasing.

    a. Use **Toggle grid and axes** to show the axes without grid.

    b. Choose **Draw** → Function. Select Type → Cartesian. In the Y = prompt type f(x).

    c. In order to modify the shape of the function to make it either strictly increasing or strictly decreasing:

        - Click on the curve, hold the mouse and drag the curve. A small circle (handle) and the name of the parameter will appear on the screen

        - Click in another part of the function and drag it.

        - There are a total of 5 different parameters that are responsible for the shape of the function, and you can change each one of them by dragging the handles.

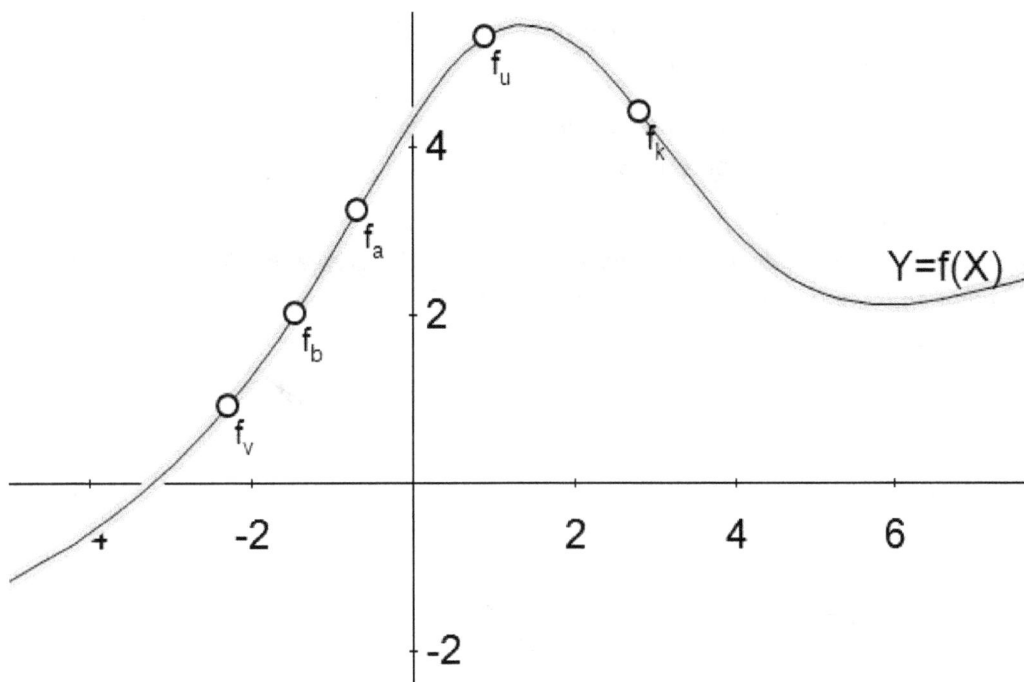

2. Draw a tangent line to the graph of *f(x)* at a point A (*t*, *f(t)*).

   a. Select the plot of the function and choose **Construct** → Tangent to Curve. The tangent line and point of tangency will be displayed (point A).

   b. Select point A and graph of *f(x)* and choose **Constrain** → Point proportional along the curve. Type *t* in the open edit box.

3. Find the derivative of *f(x)* by calculating the slope of the tangent line at point A (*t*, *f(t)*).

   a. Select the tangent line and choose **Calculate** → Slope.

Q1. How can you obtain the graph of the inverse function $f^{-1}(x)$?

A1: *The graph of the inverse function is the reflection of the graph of f(x) over the identity line y = x.*

4. Follow the method you described in Q1 and graph the inverse function $f^{-1}(x)$.

   a. Choose **Draw** → Point and plot a point O at the origin of the coordinate system.

   b. Click on the graph of the function and point A, choose **Construct** → Reflection.

c. Click on the point O and then anywhere else in the plane to graph the axis of reflection through the origin.

d. Click on the line and choose **Constrain** → Slope. Type 1 in the open edit box. Change label of point A' to B.

Note: Since the axis of reflection must be a geometric object, the line y = x must be constructed geometrically, not as a function. You can also create the line before you do the reflection and simply select the line at the axis of reflection prompt. The suggested method is one of many possible methods that students can choose.

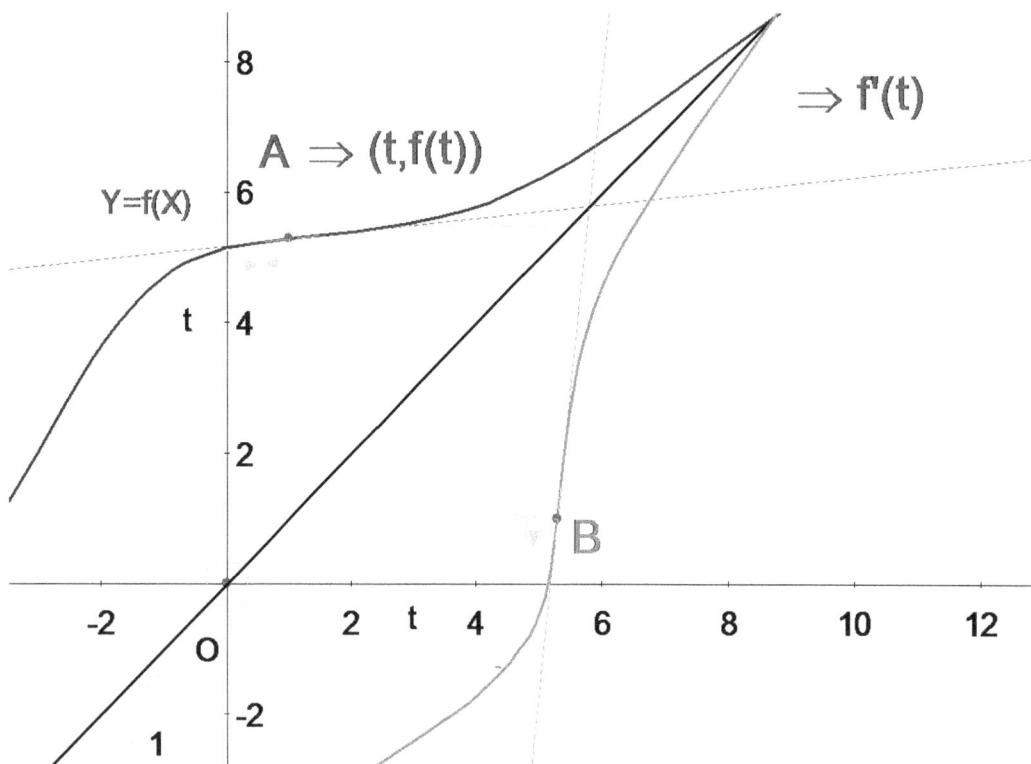

Q2. What are the coordinates of point B on the graph of $f^{-1}(x)$ that corresponds to the point A on the graph of $f(x)$?

A2: The coordinates are (f(t), t).

5.   Draw a tangent line to the graph of $f^{-1}(x)$ at point B.

a. You must delete point B (the reflected image) and recreate it in the usual way.

b. Select the plot of the function $f^{-1}(x)$ and choose **Construct** → Tangent to Curve. The tangent line and point of tangency will be displayed (label it point B).

c. Select point B and parabola and choose **Constrain** → Proportional along curve. Type *t* in the open edit box.

Note: since the inverse function's graph is a reflection of the graph of function f(x) and x and y are switched, the Point proportional along curve parameter is the y-coordinate of point B.

6. Find the derivative of $f^{-1}(x)$ by calculating the slope of the tangent line at point B.

a. Select the tangent line and choose **Calculate** → Slope

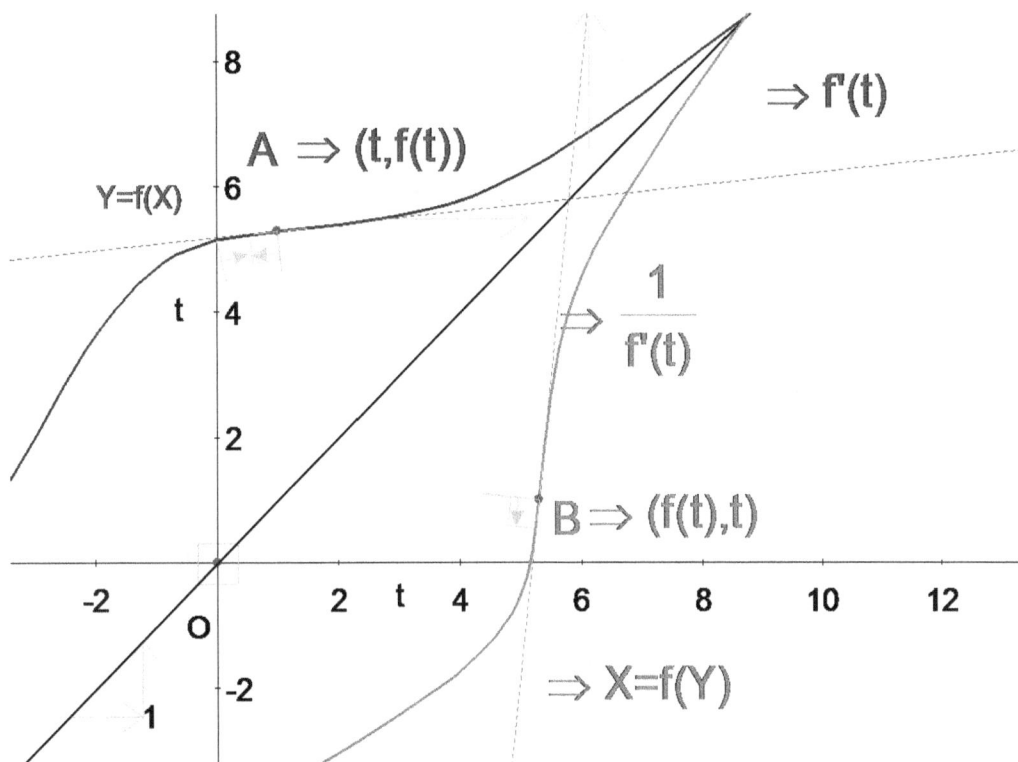

Q3. What is the relationship between the derivative of $f(x)$ and the derivative of $f^{-1}(x)$ in corresponding points?

A3: Based on results from the software, students should see that

$$\left(f^{-1}(y)\right)' = \frac{1}{f'(x)}, \text{ where } y = f(x).$$

THE SPECIAL CASE $y = \sqrt{x}$

7.   Draw the function $y = \sqrt{x}$.

    a. Choose **Draw** → Function. Select Type → Cartesian. In the Y = prompt type sqrt(x).

Q4. What is the derivative of this function?

A4: $\left(\sqrt{x}\right)' = \left(x^{1/2}\right)' = \frac{1}{2}x^{-1/2} = \frac{1}{2\sqrt{x}}$

8.   Verify your calculations using *Geometry Expressions.*

    a. Select the plot of the function and choose **Construct** → Tangent. The tangent line and point of tangency will be displayed (point A).

    b. Select point A and parabola and choose **Constrain** → Point proportional along curve. Type *a* in the open edit box.

    c. Select the tangent line and choose **Calculate** → Slope.

Q5. What is the inverse function for $y = \sqrt{x}$.

A5: $y = x^2, x \geq 0.$

9.   Complete the same steps as you used in the general case to construct the graph of the inverse function to $y = \sqrt{x}$.

Note: students should complete step 4 above.

10. Verify your formula for the inverse function using *Gx.*

    a. Click the graph of the inverse function and choose **Calculate** → Implicit equation.

Q6. What is the derivative of the inverse function at $x = a$?

A6: *Based on the relationship between derivatives* $\left(f^{-1}(y)\right)' = \dfrac{1}{f'(x)} = \dfrac{1}{1/2\sqrt{a}} = 2\sqrt{a}$.

Differentiating the inverse function directly: $(f^{-1}(x))' = (x^2)' = 2x$ *and* $x = \sqrt{y}\big|_{y=a} = \sqrt{a}$, *so we get the same result.*

11. Find the derivative of the inverse function at a corresponding point using Gx.

   a. Select the plot of the inverse function and choose **Construct** → Tangent to Curve. The tangent line and the point of tangency will be displayed (point B).

   b. Select point B and the graph of the inverse function and choose **Constrain** → Point proportional along the curve. Type *a* in the open edit box.

   c. Select the tangent line and choose **Calculate** → Slope.

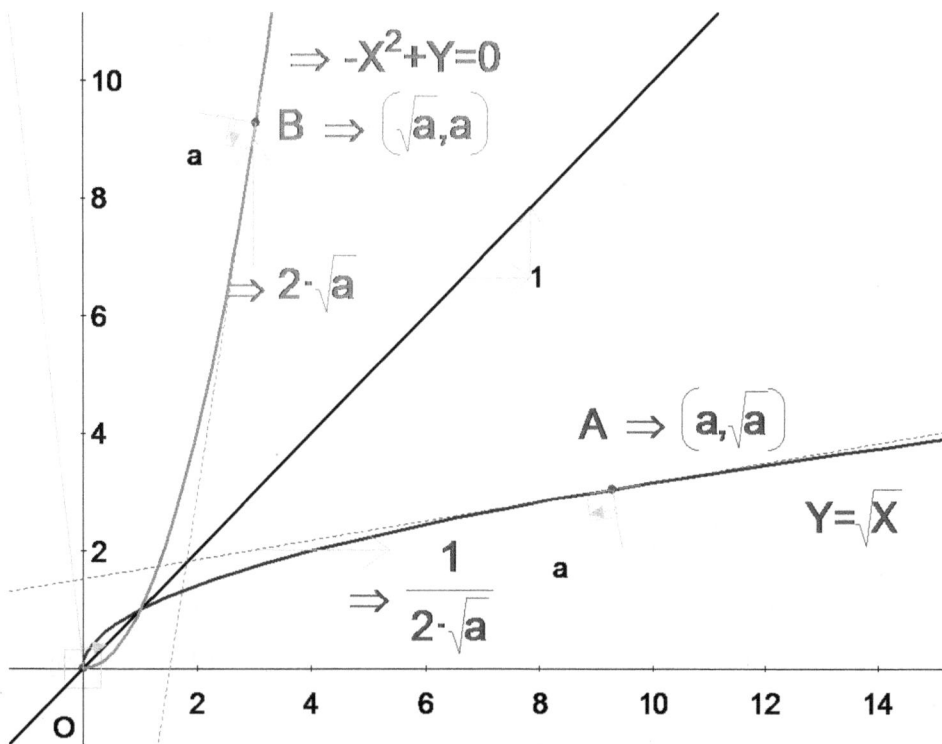

Derivative of an Inverse Function

Exploration 2.6: Given a differentiable and monotonic function, *f(x)*.   What is the relationship between the derivative of the function and its inverse?

INVESTIGATION: THE GENERAL CASE

1.   Draw a general function $y = f(x)$. Drag it at several places to make it either strictly increasing or strictly decreasing.

2.   Draw a tangent line to the graph of $f(x)$ at a point A $(t, f(t))$.

3.   Find the derivative of $f(x)$ by calculating the slope of the tangent line at point A $(t, f(t))$.

Q1.   How can you obtain the graph of the inverse function $f^{-1}(x)$?

4.   Follow the method you described in Q1 and graph the inverse function $f^{-1}(x)$.

Q2.   What are the coordinates of the point B on the graph of $f^{-1}(x)$ that corresponds to the point A on the graph of $f(x)$?

5.   Draw a tangent line to the graph of $f^{-1}(x)$ at a point B.

6.   Find the derivative of $f^{-1}(x)$ by calculating the slope of the tangent line at point B.

Q3.   What is the relationship between the derivative of $f(x)$ and the derivative of $f^{-1}(x)$ in corresponding points?

THE SPECIAL CASE $y = \sqrt{x}$

7.   Draw the function $y = \sqrt{x}$.

Q4.   What is the derivative of this function?

8.   Verify your calculations using *Gx*.

Q5.   What is the inverse function for $y = \sqrt{x}$.

9.   Complete the same steps as you used in the general case to construct the graph of the inverse function to $y = \sqrt{x}$.

10.   Verify your formula for the inverse function using *Gx*.

Q6.   What is the derivative of the inverse function $x = a$?

11. Find the derivative of the inverse function at a corresponding point using Gx.

3. Applications of Derivatives

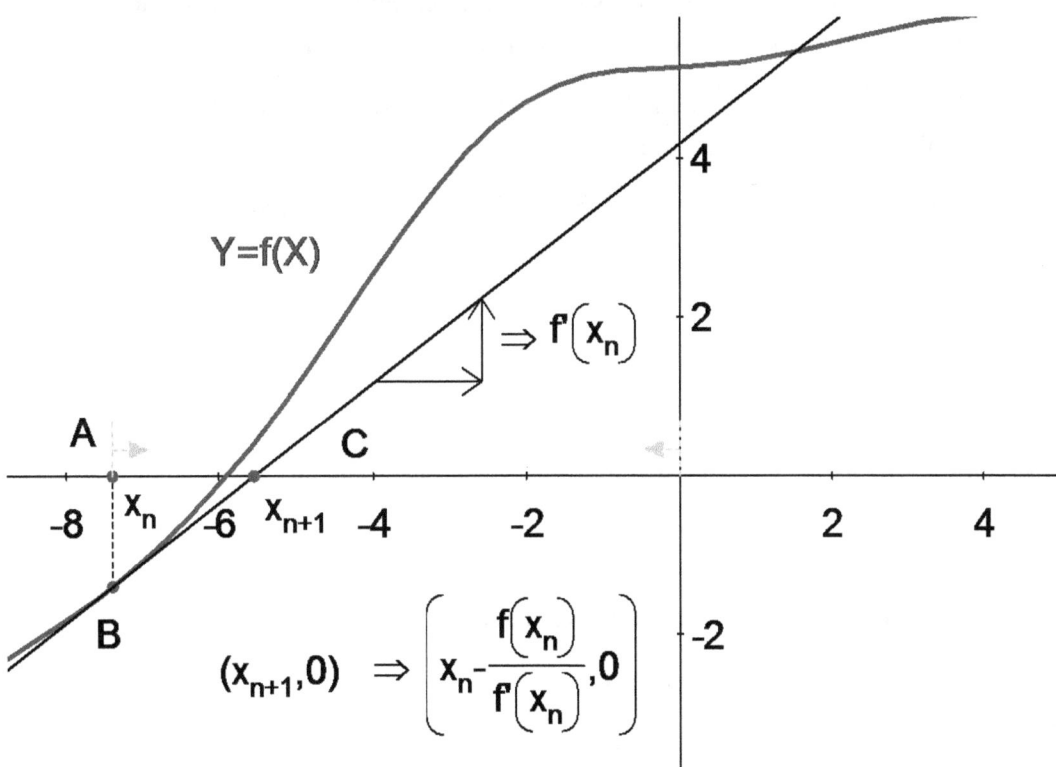

$$Y=f(X)$$

$$\Rightarrow f'\left(x_n\right)$$

$$(x_{n+1},0) \quad \Rightarrow \quad \left(x_n - \frac{f\left(x_n\right)}{f'\left(x_n\right)}, 0 \right)$$

3.1 Envelope of a Parabola

Exploration 3.1: A tangent line is drawn to the graph of a function $y = x^2$. What is the set of all points that are swept by the tangent line as it moves along the graph of the function?

SUMMARY

Mathematics Objectives:

- Explore the plane region that is swept by the tangent line as it moves along the function curve (also known as the envelope of the function).

- Determine if the tangent line to the graph of the given function can pass through an arbitrary point on the coordinate plane.

- Determine relationship between the envelope of a function and its concavity.

Vocabulary:

- Envelope of a parabola

- Concavity of a function

- Locus

Pre-requisites:

- Quadratic formula.

- Slope-intercept form of the equation of a line.

- Derivative as a slope of a tangent line.

Problem Notes:

- In geometry students study the locus. A locus in a plane is the set of all points in the plane that satisfy a given condition or a set of given conditions. In this problem students explore the locus of points in a plane that is swept by the tangent line as it moves along the function curve. In mathematics this locus is referred to as the envelope of the function.

- The location of the envelope of a function in relation to the graph of a function determines the concavity of a function. If an envelope of a function is below the graph of a function on a given interval, the function is concave up on this interval. If an envelope of a function is above the graph of a function on a given interval, the function is concave down on this interval.

- A parabola separates the plane into three distinct sets of points: those on the graph of the function, those in the interior and those in the exterior of the graph of the function. If you choose a point in the plane that is not on the function, is there a tangent line to the function containing the point? The symbolic features of *Gx* enable students to investigate this question.

Technology skills:

- Draw: function, point

- Construct: tangent line to a curve, trace

- Constrain: a point proportional along the curve, coordinates of a point

- Calculate: implicit equation

Extension:

Find the equation of a tangent line to $y = x^2$ that intersects the *x*-axis at an angle α.

STEP-BY-STEP INSTRUCTIONS

CONSTRUCTION

1. Draw the function $y = x^2$. The plot of the parabola will appear on the screen.

   a. Choose **Draw** $\rightarrow$ Function.

   b. Choose Cartesian for the Type of the function. In the Y= prompt type x^2. Then press OK.

2. Construct a tangent line to the curve. Constrain the x – coordinate of the point of tangency to be *a*.

   a. Select the graph of the function and choose **Construct** $\rightarrow$ Tangent Line.

    b. Select point A and the graph of the function and choose **Constrain** → Point proportional along curve.

    c. Type *a* in the open edit box.

Q1. Write the equation of the tangent line.

A1: *The slope of the tangent line is equal to the derivative of the function at x = a, so m = 2a. Use the slope-intercept form of an equation for the tangent line to determine the y-intercept:* $a^2 = 2a(a) + b$, *so b = -a^2. The final equation:* $y = 2ax - a^2$

3. Use *Geometry Expressions* to verify your equation.

    a. Select the tangent line and choose **Calculate** → Implicit Equation.

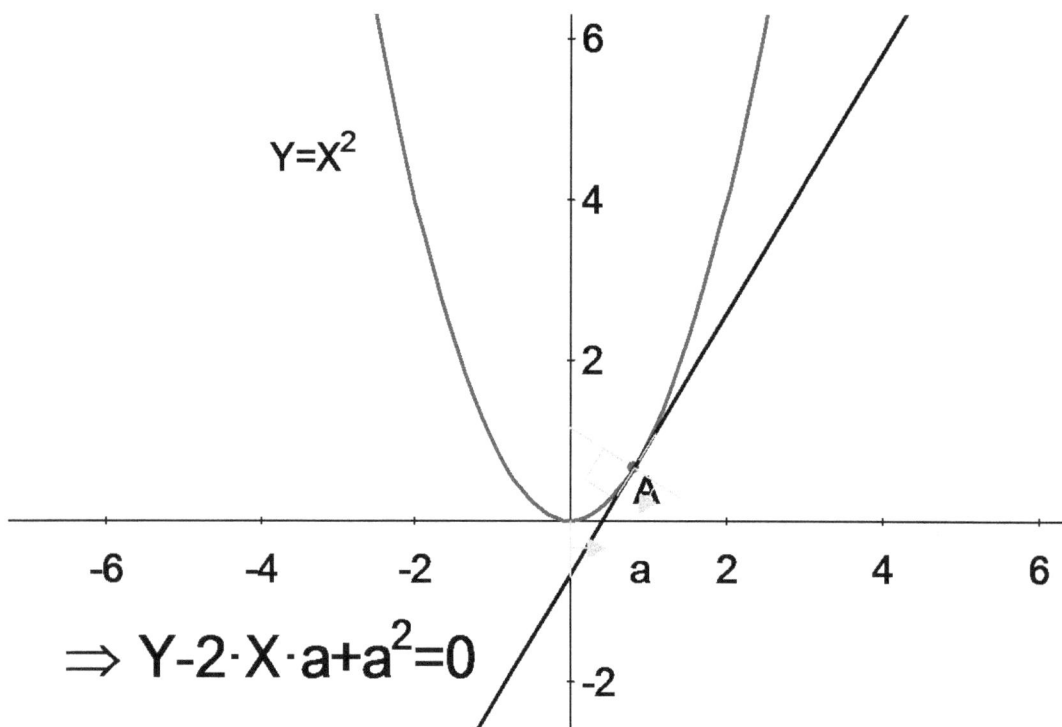

$$Y = X^2$$

$$\Rightarrow Y - 2 \cdot X \cdot a + a^2 = 0$$

INVESTIGATION

4.   Draw point B and constrain its coordinates to (x_b, y_b).

   a.   Choose **Draw** → Point.

   b.   Select the point and choose **Constrain** → Coordinates. In the open edit box type $x[b], y[b]$.

5.   Drag point A to see if the tangent line will pass through point B for various locations of point B.

Note: students can drag point A to see if the tangent line passes through point B. Then they can move point B to a different location and adjust the position of point A to see if the tangent line passes through point B. They should repeat this process until they can formulate their conjecture.

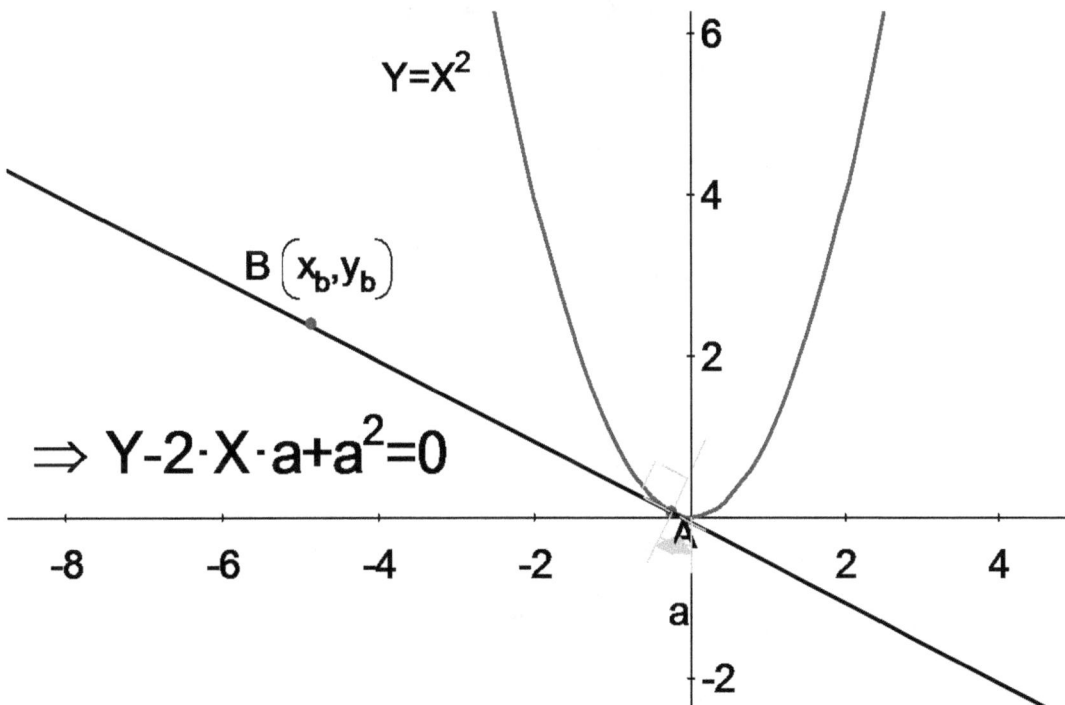

Q2.   Can the tangent line to the graph of the function $y = x^2$ pass through an arbitrary point on the coordinate plane?

A2: *The parabola divides the plane into two regions, the set of points that the tangent line cannot pass through in the interior of the parabola where $y > x^2$, and set of points that the tangent line passes through that include points on the parabola and in the exterior of the parabola, and where $y \leq x^2$.*

Q3. Provide calculations that support your answer.

A3: *Substitute the coordinates of point B into the tangent line equation and analyze the solution of the equation for the value of a. $y_b - 2x_b a + a^2 = 0$. Using the quadratic formula: $a = x_b \pm \sqrt{x_b^2 - y_b}$. In the region above the parabola $y > x^2$, so there are no solutions. In the region below the parabola $y < x^2$, so there are 2 solutions. For all points on the parabola, $y = x^2$, there is only one solution, the point of tangency.*

6.   Trace the tangent line to observe the envelope of the function $y = x^2$.

   a.   Select tangent line and choose **Construct** → Trace.

   b.   In the open window choose the parametric value to be *a*, Start Value -10, End Value 10, Count 100.

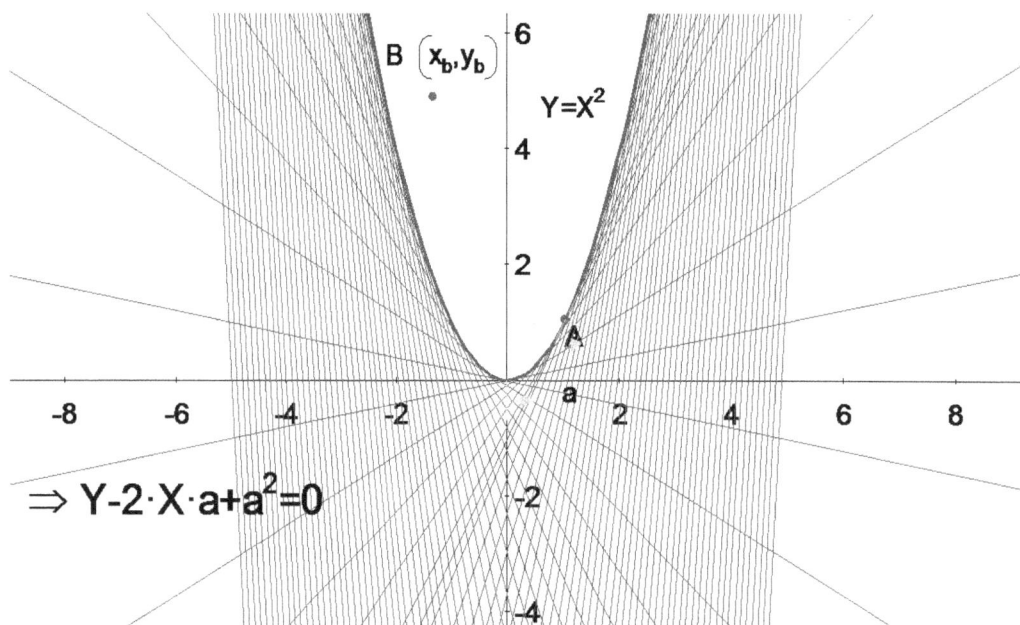

Q4. Does the software support your conclusion about the locations of all points that are swept by the tangent line?

A4: *The region that is swept by the tangent line includes all the points under the parabola and the parabola itself.*

Q5. How do your observations relate to the concavity of the function $y = x^2$?

A5: *The tangent line to the parabola is always under the parabola, representing that $y = x^2$ is concave up.*

Envelope of a Parabola

<u>Exploration 3.1:</u> A tangent line is drawn to the graph of a function $y = x^2$. What is the set of all points that are swept by the tangent line as it moves along the graph of the function?

CONSTRUCTION

1. Draw function $y = x^2$.

2. Construct a tangent line to the curve. Constrain the x – coordinate of the point of tangency to be a.

Q1. Write an equation of the tangent line.

3. Use Gx to verify your equation.

INVESTIGATION

4. Draw point B and constrain its coordinates to (x_b, y_b).

5. Drag point A to see if the tangent line will pass through point B for various locations of point B.

Q2. Can the tangent line to the graph of the function $y = x^2$ pass through an arbitrary point on the coordinate plane?

Q3. Provide calculations that support your answer.

6. Trace the tangent line to observe the envelope of the function $y = x^2$.

Q4. Does the software support your conclusion about locations of all points that are swept by the tangent line?

Q5. How do your observations relate to the concavity of the function $y = x^2$?

3.2 Linear Approximation

Exploration 3.2: Explore the linear approximation of a function $y = f(x)$ at a given point, if a function is differentiable on an interval containing this point.

SUMMARY

Mathematics Objectives:

- Use a tangent line to approximate a value of a function at a given point.

- Explore factors that affect the precision of linear approximation.

Vocabulary:

- Linear approximation

- Precision

- Over- and under-estimation

Pre-requisites:

- Differentiability of a function

- Derivative as a slope of the tangent line

- Point-slope form of an equation of a line

Problem Notes:

- Students do not necessarily see the importance of this particular application, since in their minds "exact" values of a function can be found using technology (calculators, computers, even cell phones). This exploration, however, is important from both theoretical and practical perspectives.

- This method is an introduction to the more general concept of local linearity used in practically all aspects of differential and integral calculus. More specifically, for any differentiable function it is true that if you are sufficiently close to a point $(a, f(a))$ then the change in the function is approximately the same as the change in the tangent line to the graph of the function at the point $(a, f(a))$ for the same Δx, i.e.

$\Delta y \approx dy$. Thus, the tangent line provides a good approximation of the function on the interval $(a, a + \Delta x)$.

- This exploration provides students with the practical understanding of how computational devices evaluate functions using various approximation techniques.

- Students first consider a generic differentiable function $y = f(x)$. They construct the tangent line to the graph of the function at a given point, derive the equation for the tangent line, and use the tangent line to find the approximate value of the function at a different point. Thus, students derive a general formula for linear approximation of a function using tangent lines.

- Students then explore the difference between the actual and approximate values of a function to explore factors that affect the precision of their approximation. They also explore how the position of the tangent line in relationship to the graph of the function causes the approximated value to be an over- or under- estimation.

- As an application, students use the linear approximation method for three specific functions, where they also determine the precision of an approximation, and if it is an over- or under-estimation.

Technology skills:

- Draw: function, line segment

- Constrain: a point proportional along the curve, coordinates of a point

- Construct: tangent line, perpendicular, intersection

- Calculate: coordinates of a point, slope, distance / length

Extension:

Find the piecewise linear approximation of the function $y = e^x$ on an interval [0, 1] so that the approximated value of the function for any point on this interval has a precision of 0.1 or less. What is the least number of linear segments that are necessary for this approximation?

STEP-BY-STEP INSTRUCTIONS

LINEAR APPROXIMATION

1. Draw a function $y = f(x)$. Modify the shape of the function as needed.

    a. Use **Toggle grid and axes** to show the axes without grid.

    b. Choose **Draw** → Function. Select Type → Cartesian. In the Y = prompt type f(x).

    c. In order to modify the shape of the function:

        1) Click on the curve, hold the mouse and drag the curve. A small circle (handle) and the name of the parameter will appear on the screen.

        2) Click in another part of the function and drag it.

        3) There are total of 5 different parameters that are responsible for the shape of the function, and you can change each one of them by dragging the handles.

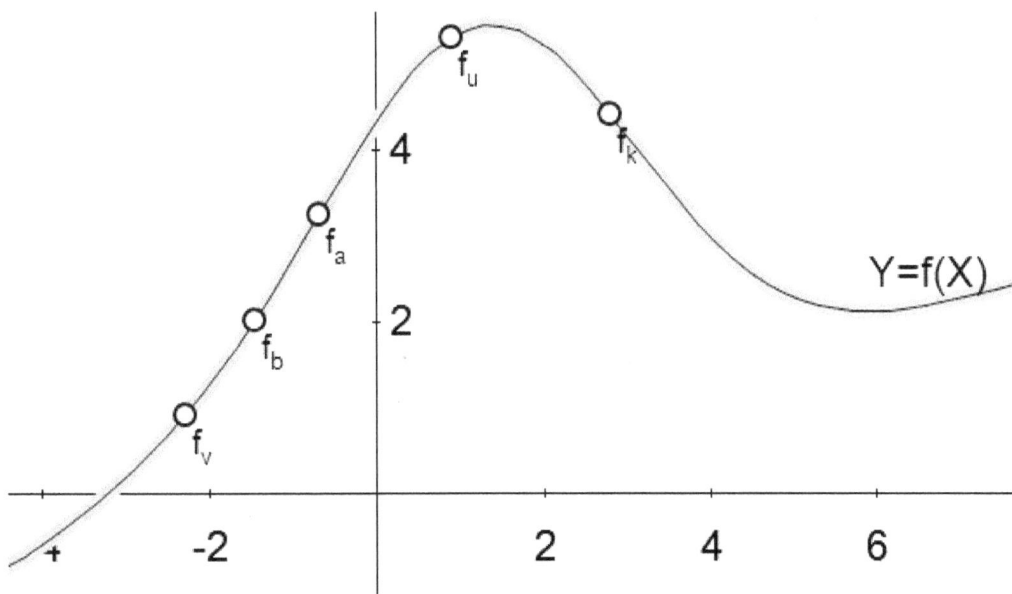

2. Draw a point $x = x_0$ on the x-axis. Then draw a tangent line to the graph of the function at a point $x = x_0$ where $f(x)$ is differentiable.

    a. Choose **Draw** → Point to draw point A on the x − axis.

c. Select point A and the *x*-axis, and choose **Constrain** → Proportional along the curve. Type x[0] in the open edit box.

d. Select the plot of the function and choose **Construct** → Tangent to Curve. The tangent line and the point of tangency will be displayed (point B).

e. Select point B and the graph of *f(x)* and choose **Constrain** → Point proportional along curve. Type x[0] in the open edit box.

f. Connect points A and B with a line segment by choosing **Draw** → Line Segment

Q1. What is the value of the function and slope of the tangent line at $x = x_0$?

A1: The value of the function is f(x₀) and the slope of the tangent line is f'(x₀).

3. Verify your results using the software.

a. Select the point B and choose **Calculate** → Coordinates

b. Select the tangent line and choose **Calculate** → Slope.

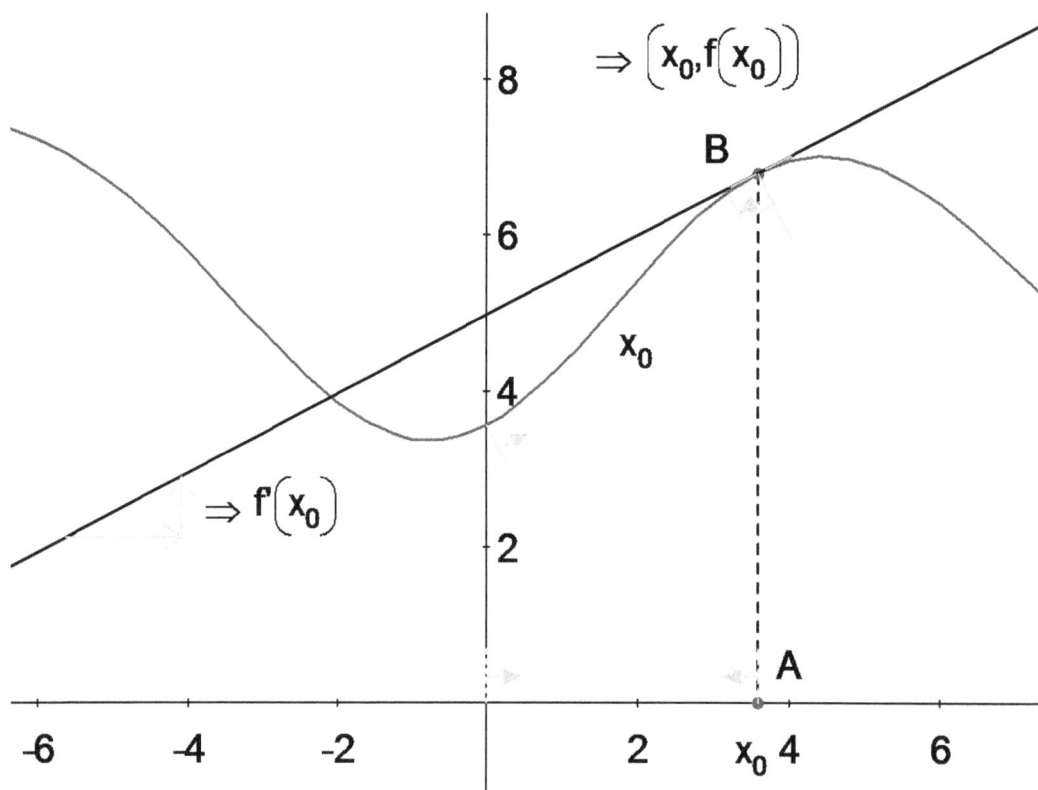

Q2. Using the tangent line to the graph of $f(x)$ through point B, find the approximate value of the function at $x = x_0 + t$.

A2: Students should derive an equation of the tangent line and then substitute $x = x_0 + t$ into this equation. The equation of the tangent line is: $y = f'(x_0)(x - x_0) + f(x_0)$. After substitution, $f(x_0 + t) \approx f(x_0) + t \cdot f'(x_0)$.

4. Verify your calculations using the software.

   a. Draw point C on the x-axis with the coordinates $x = x_0 + t$ by choosing **Draw** → Point. Select point C and the x-axis, and choose **Constrain** → Proportional along the curve. Type x[0]+t in the open edit box.

   b. Select point C and the x-axis and choose **Construct** → Perpendicular.

   c. Select the perpendicular line and the tangent line and choose **Construct** → Intersection. The point of intersection is D.

   d. Click point D and choose **Calculate** → Coordinates.

Q3. How close is your approximation to the actual value of the function?

A3: The absolute value of the difference between the actual and approximate values of the function is the distance between point D and the point of intersection between the perpendicular and the graph of f(x). Students can drag point C to change the value of t and discuss qualitatively how close the approximated value is to the exact value.

5. Find the difference between the actual and approximate values of $f(x)$ at $x = x_0 + t$ using the software.

   a. Choose **Draw** → Point and draw an arbitrary point E (in the plane and not on the graph of function or coordinate axes).

   b. Select the point and choose **Constrain** → Coordinates. In the open edit box type (x[0]+t,f(x[0]+t)). This will put the point E on the graph of $f(x)$.

   c. Choose **Draw** → Line Segment and connect points D and E.

   d. Select the segment and choose **Calculate** → Distance.

*Note: In order to have an uncluttered diagram, students can hide all unnecessary information by selecting it first and then choosing **View** → Hide (or Ctrl – H)*

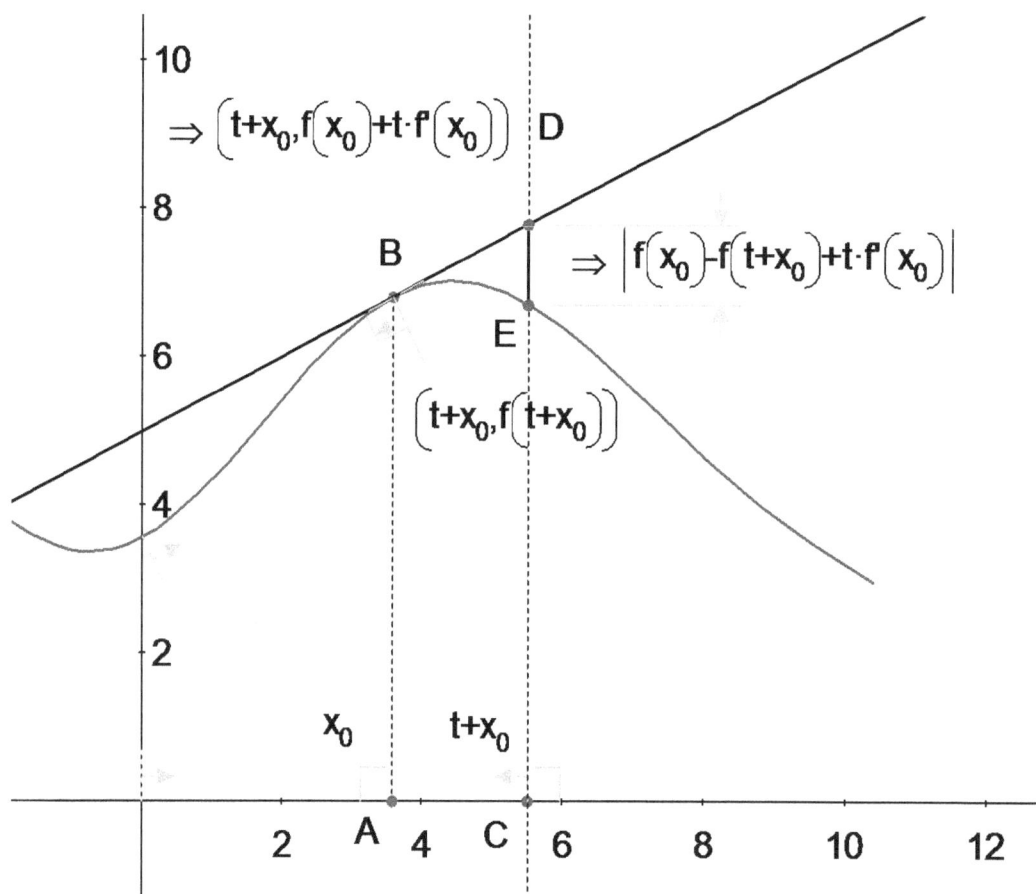

The graph shows a curve and a tangent line with the following labels:

- $\Rightarrow \left(t+x_0, f\left(x_0\right)+t\cdot f'\left(x_0\right)\right)$ D
- B
- $\Rightarrow \left|f\left(x_0\right)-f\left(t+x_0\right)+t\cdot f'\left(x_0\right)\right|$
- E
- $\left(t+x_0, f\left(t+x_0\right)\right)$
- x_0
- $t+x_0$
- A
- C

6. Graph the difference between the actual and approximate values of $f(x)$ (not the distance) **as a function of** t. (Hide all calculations except for the difference).

   a. Select the expressions for the calculated difference and choose **Edit** → Copy As → String

   b. Choose **Draw** → Function and use Ctrl-V to paste the expression for the difference into Y=.

   c. Before you hit enter you need to delete the "abs" in the pasted expression in order to graph the actual difference and not the absolute value of it. You will also need to replace t with X in this expression.

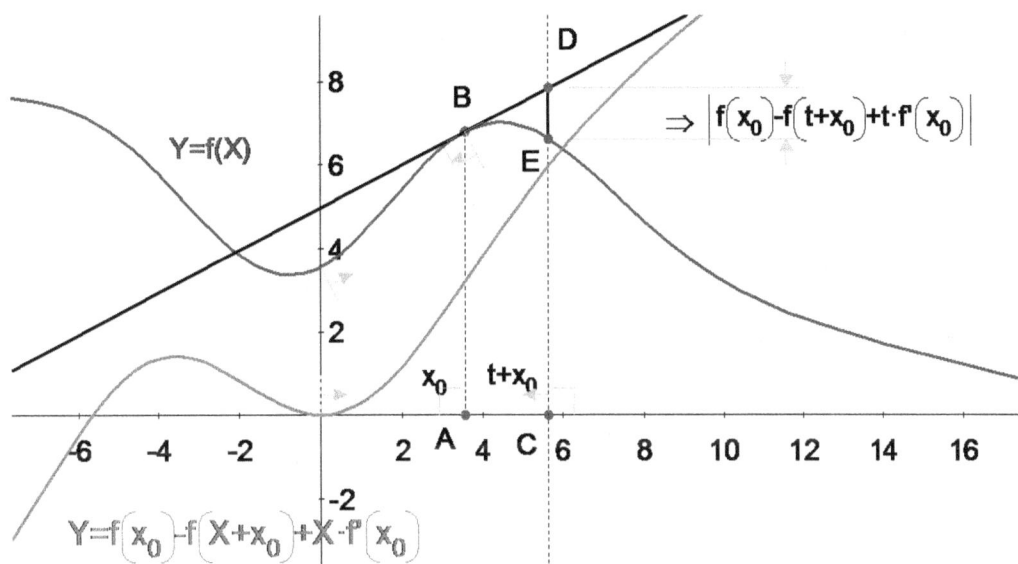

The graph shows $Y=f(X)$, a line through points B, D, E with the tangent line equation $Y=f\left(x_0\right)-f\left(X+x_0\right)+X\cdot f'\left(x_0\right)$, labeled points B, D, E, A, C, with x_0 and $t+x_0$ marked on the x-axis, and the expression $\Rightarrow \left|f\left(x_0\right)-f\left(t+x_0\right)+t\cdot f'\left(x_0\right)\right|$.

Q4. What factors affect the precision of the linear approximation?

A4: Students' answers will vary. They are expected to state that points B and C should be relatively close, so that the function is more "linear" on the interval where the tangent line is used for the approximation.

Note: In order to see the difference between the actual and approximated values of the function on the graph while dragging point C along the x-axis, students can plot point T on the graph that corresponds to a specific value of t. Here are the steps of construction:

a. Choose **Draw** → Point and draw an arbitrary point T (in the plane not on the graph of function or coordinate axes).

b. Select the point and choose **Constrain** → Coordinates. In the open edit box type $(t, f(x[0]) - f(x[0]+t) + t*f'(x[0]))$.

c. Drag the point C and observe the point T. The precision of approximation is determined by how close the point T to the *x*-axis.

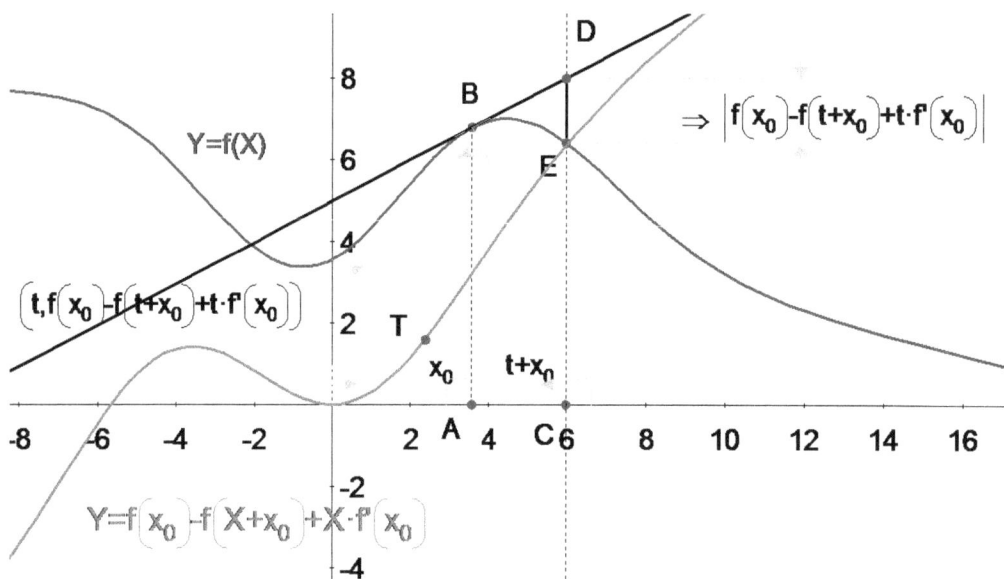

The graph shows $Y=f(X)$, a tangent line, with points A, B, C, D, E, T labeled.

$$\Rightarrow \left| f\left(x_0\right) - f\left(t + x_0\right) + t \cdot f'\left(x_0\right) \right|$$

$$\left(t, f\left(x_0\right) - f\left(t + x_0\right) + t \cdot f'\left(x_0\right) \right)$$

$$Y = f\left(x_0\right) - f\left(X + x_0\right) + X \cdot f'\left(x_0\right)$$

Q5. How do you know when your approximation is an over-estimation? An under-estimation?

A5: *When the tangent line is above the graph of the function, it is an over-estimation. In this case point T is above the x-axis. When the tangent line is under the graph of the function, it is an under-estimation. In this case point T is below the x-axis.*

Note: Students should drag point A and observe how the relative position of the graph of f(x) and the tangent line at point A affect the precision of the approximate value.

APPLICATIONS

Q6. Find an equation of the tangent line to the function $y = f(x)$ at a point A. Use linear approximation to estimate the value of the function at points B and C. Determine if it under or over estimates in each case. Confirm your calculations using the software.

Function	Point A	Point B	Point C
$f(x) = \sqrt{x}$	$x = 4$	$x = 3.8$	$x = 4.2$
$f(x) = \sin x$	$x = \pi$	$x = 3$	$x = 3.3$
$f(x) = e^x$	$x = 0$	$x = -0.1$	$x = 0.1$

A6:

1) $f(x) = \sqrt{x}$, $f'(x) = \dfrac{1}{2\sqrt{x}}$. *So f(4)=2,* $f'(4) = \dfrac{1}{4}$.

Point B: $\sqrt{4-0.2} \approx 2 + (-0.2) \cdot \dfrac{1}{4} = 1.95$. *Point C:* $\sqrt{4+0.2} \approx 2 + 0.2 \cdot \dfrac{1}{4} = 2.05$.

   a. Choose **Draw** → Function. In the open edit box choose Cartesian for the Type and in Y= prompt type sqrt(x).

   b. Select the plot of the function and choose **Construct** → Tangent to Curve. The tangent line and point of tangency will be displayed (point A).

   c. Select point A and the curve, and choose **Constrain** → Proportional along the curve. In the open edit box type 4.

   d. Draw points B and C on the curve.

   e. Repeat step c. for the points B and C, typing corresponding *x* values.

   f. Zoom in (+) to see the points separately.

   g. Select the tangent line and choose **Calculate** → Implicit Equation.

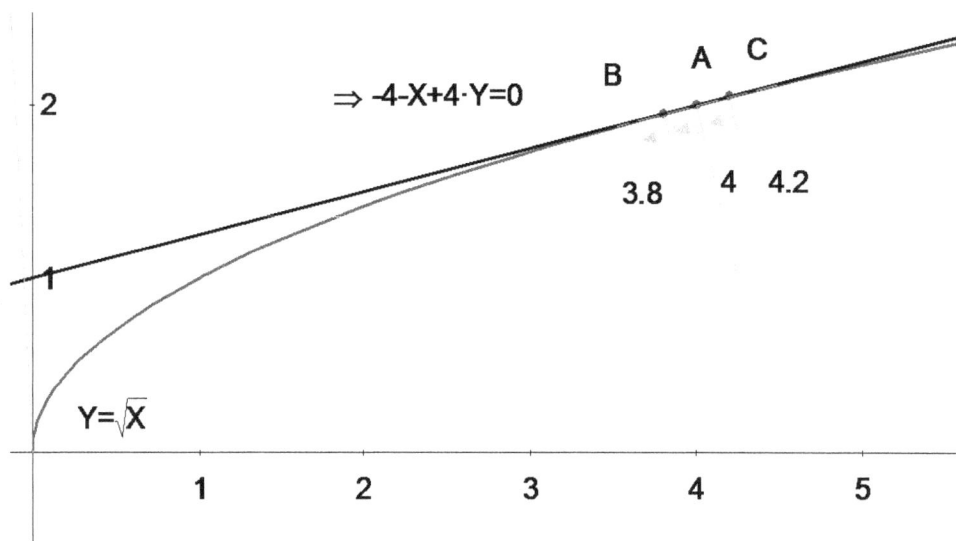

Students can substitute values of x for points B and C into the equation and confirm their results. Both approximations are over-estimations since the tangent line is above the graph of the function.

2) $f(x) = \sin x$, $f'(x) = \cos x$. So $f(\pi)=0$, $f'(\pi)=-1$.

Point   B: $\sin(3) \approx 0 + (3-\pi)\cdot(-1) = \pi - 3 = 0.14$.   *Point   C:*
$\sin(3.3) \approx 0 + (3.3-\pi)\cdot(-1) = -0.16$.

Complete the constructions by following the steps above. The approximation at point B is an over-estimation since the tangent line is above the curve, and the approximation at point C is an under-estimation since the tangent line is below the graph of the function.

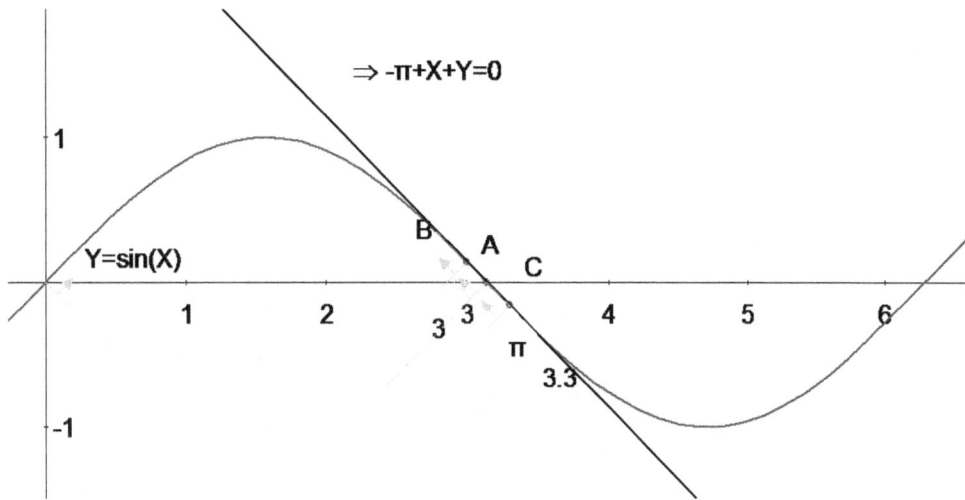

$$\Rightarrow -\pi + X + Y = 0$$

Y=sin(X)

3) $f(x) = e^x$, $f'(x) = e^x$. So $f(0)=1$, $f'(0)=1$.

Point B: $e^{-0.1} \approx 1 + (-0.1) = 0.9$. *Point C:* $e^{0.1} \approx 1 + (0.1) = 1.1$.

Complete the constructions by following the steps above.   Both the approximations are under-estimations since the tangent line is under the graph of the function.

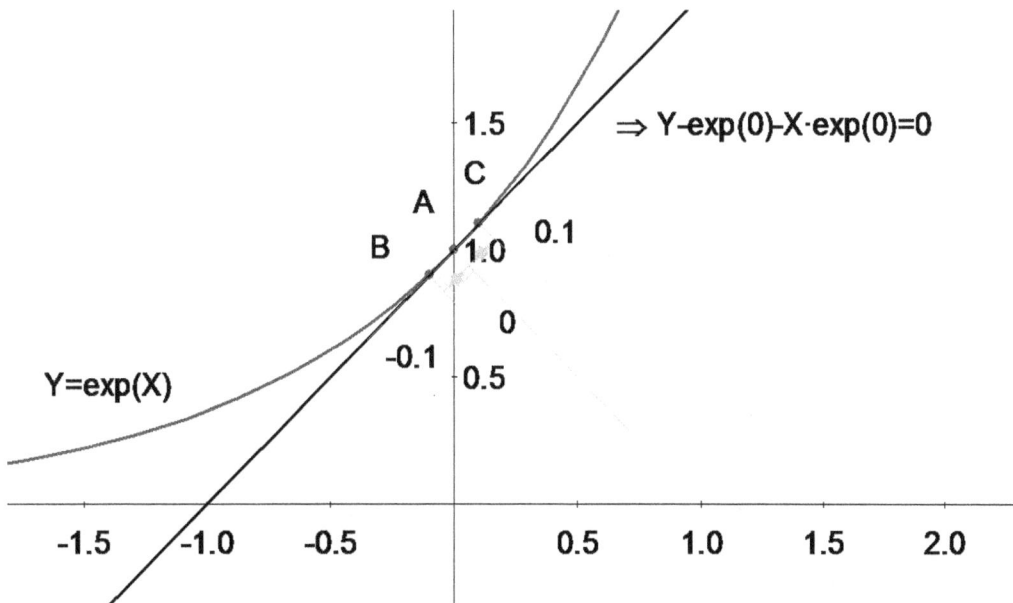

$$\Rightarrow Y - \exp(0) - X \cdot \exp(0) = 0$$

Y=exp(X)

Linear Approximation

Exploration 3.2: Explore the linear approximation of a function $y = f(x)$ at a given point, if a function is differentiable on an interval containing this point.

LINEAR APPROXIMATION

1.   Draw a function $y = f(x)$. Modify the shape of the function as needed.

2.   Draw a point $x = x_0$ on the x-axis. Then draw a tangent line to the graph of the function at a point $x = x_0$ where $f(x)$ is differentiable.

Q1.   What is the value of the function and slope of the tangent line at $x = x_0$?

3.   Verify your results using the software.

Q2.   Using the tangent line to the graph of $f(x)$ through the point B find the approximate value of the function at $x = x_0 + t$.

4.   Verify your calculations using the software.

Q3.   How close is your approximation to the actual value of the function?

5.   Find the difference between the actual and approximate values of $f(x)$ at $x = x_0 + t$ using the software.

6.   Graph the difference between the actual and approximate values of $f(x)$ as a function of t. (Hide all calculations except for the difference).

Q4.   What factors affect the precision of the linear approximation?

Q5.   How do you know when your approximation is an over-estimation? An under-estimation?

APPLICATIONS

Q6. Find an equation of the tangent line to the function $y = f(x)$ at the point A. Use linear approximation to estimate the value of the function at points B and C. Determine if it under or over estimates in each case. Confirm your calculations using the software.

Function	Point A	Point B	Point C
$f(x) = \sqrt{x}$	$x = 4$	$x = 3.8$	$x = 4.2$
$f(x) = \sin x$	$x = \pi$	$x = 3$	$x = 3.3$
$f(x) = e^x$	$x = 0$	$x = -0.1$	$x = 0.1$

3.3 Newton's Method

<u>Exploration 3.3:</u> In this activity you will explore Newton's method. Consider a function $y = f(x)$ such that $f(c) = 0$. Start with an initial guess which is reasonably close to the true root and then approximate the function by its tangent line. Compute the x-intercept of this tangent line. This x-intercept will typically be a better approximation to the function's root than the original guess, and the method can be iterated.

SUMMARY

<u>Mathematics Objectives:</u>

- To approximate the zeros of a function using the iteration technique known as Newton's Method.

- To determine when Newton's method fails to produce a zero.

<u>Vocabulary:</u>

- Iteration

- Seed value

<u>Pre-requisites:</u>

- Point-slope form of the equation of a line.

- Recursive formula.

- Derivative as a slope of a tangent line

- Convergence and divergence of sequences

- Differentiability and continuity of functions

<u>Problem Notes:</u>

- Isaac Newton was perhaps the most influential mathematician of his time. Newton first described his method for finding zeros in 1669 in his book *De Analysi per Aequationes Numero Terminorum Infinitas.* His iteration technique was able to produce zeros for most functions quite quickly. In the 17th century this was remarkable and most necessary.

- In this exploration students will use *Geometry Expressions* to derive a recursive formula for the iterations necessary to find the zero of a generic differentiable function $y = f(x)$, assuming that the function has a zero, *e.g.* there exists an $x = c$ in the domain of the function such that $f(c) = 0$.

- Students will also explore several examples where Newton's method fails in order to analyze the conditions on the function and the initial estimate (seed value) needed to ensure that the method works.

- It is important to note that the method may fail if

  1. the initial value is too far from the true zero

  2. the function is not differentiable

  3. the derivative is zero at a point where the function has its zero. Similarly, when the derivative is close to zero, the tangent line is nearly horizontal and hence may "shoot" way past the desired root.

Technology skills:

- Draw: function

- Constrain: a point proportional along the curve, coordinates of a point

- Construct: tangent line, intersection

- Calculate: coordinates of a point, slope

Extension:

Given $f(x) = e^x$ and $x_0 = a$, use Newton's method to find the zero of the function.

Determine the sequence of iterations x_n and analyze its convergence.

STEP-BY-STEP INSTRUCTIONS

EXPLORING NEWTON'S METHOD

1. Draw a function $y = f(x)$. Modify the shape of the function as needed to make sure that the graph of the function crosses the x-axis.

    a. Use **Toggle grid and axes** to show the axes without a grid.

    b. Choose **Draw** → Function. Select Type → Cartesian. In the Y = prompt type f(x).

---

c. In order to modify the shape of the function:

1) Click on the curve, hold the mouse and drag the curve. A small circle (handle) and the name of the parameter will appear on the screen.

2) Click in another part of the function and drag it.

3) There are total of 5 different parameters that are responsible for the shape of the function, and you can change each one of them by dragging the handles.

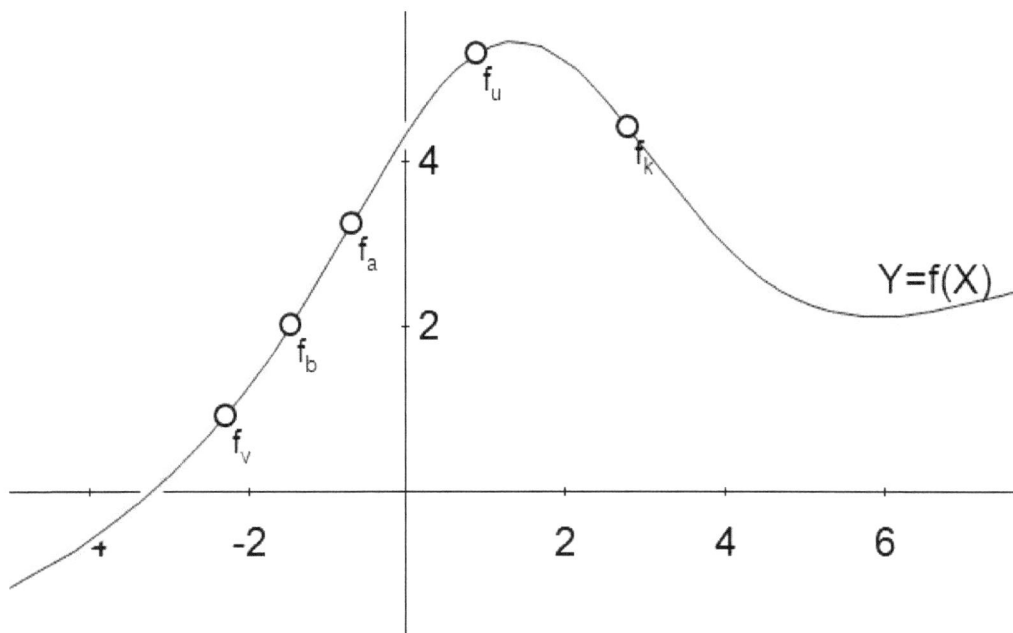

2. Let x_0 be an initial estimate for the x-intercept of the function $f(x)$ (this is also called the *seed* value). Draw this point on the x-axis.

a. Choose **Draw** → Point and plot point A on x-axis.

b. Select point A and the x-axis and choose **Constrain** → Proportional along curve.

c. Type x_0 in the open edit box.

3. Draw a tangent line to the graph of the function at a point $(x_0, f(x_0))$.

a. Select the plot of the function and choose **Construct** → Tangent. The tangent line and the point of tangency will be displayed (point B).

    b.   Select point B and the curve and choose **Constrain** → Point proportional along curve. Type x_0 in the open edit box.

Q1. What is the slope of the tangent line at this point?

A1: $m = f'(x_0)$

4.   Verify your result using *Gx*.

    a.   Select the tangent line and choose **Calculate** → Slope

5.   Construct an intersection of the tangent line and the *x*-axis. This point is your 1st iteration. Call it x_1.

    a.   Select the tangent line and *x*-axis and choose **Construct** → Intersection. The point of intersection is C.

Q2. Which is the better approximation for the *x*-intercept of $f(x)$, x_0 or x_1? Why?

A2: Answers may vary depending on the function and the choice of x_0. However, it is expected that all students base their answers on the proximity of x_0 or x_1 to the x-intercept.

Q3. How did you determine x_1?

A3: Find equation of the tangent line and set y=0 and solve for x. The equation of the tangent line is $y = f(x_0) + f'(x_0)(x_1 - x_0) = 0$, so $x_1 = x_0 - \dfrac{f(x_0)}{f'(x_0)}$.

6.   Use the software to verify your equation.

    a.   Click on the point C and choose **Calculate** → Coordinates.

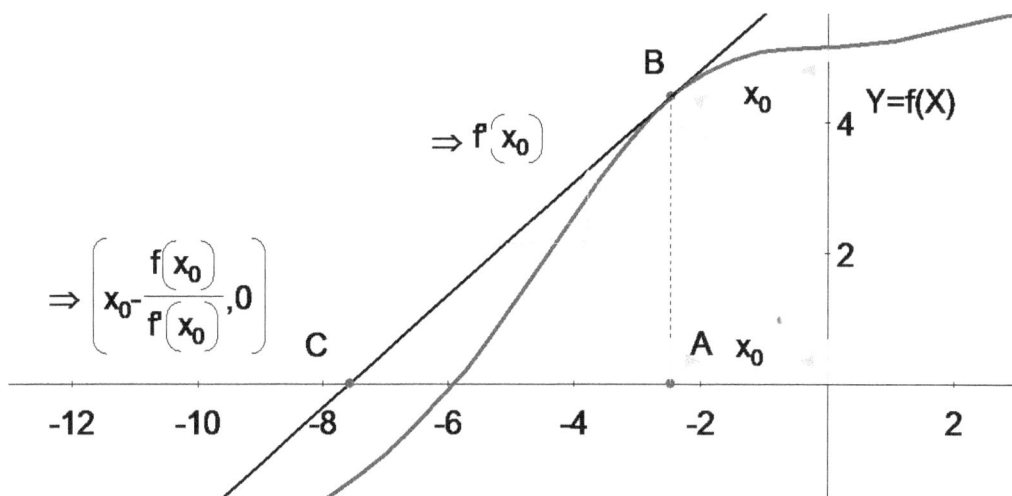

Q4. Repeat the same process, starting with the 1st iteration, x_1 to find 2nd iteration, x_2. What is x_2 in terms of x_1?

A4: *Since the procedure is the same, the formula is the same:* $x_2 = x_1 - \dfrac{f(x_1)}{f'(x_1)}$

7. Complete this process graphically and construct the point x_2 on the x-axis.

    a. Delete the calculated coordinates for the point C. Click on point C and choose **Annotate** → Coordinates. In the open edit box type (x[1],0).

    b. Select the plot of the function and choose **Construct** → Tangent to Curve. The 2nd tangent line and point of tangency will be displayed (point D).

    c. Select point D and the curve and choose **Constrain** → Proportional along the curve. In the open edit box type: x[0]-f(x[0])/f'(x[0]).

    d. Select this tangent line and the x-axis, and choose **Construct** → Intersection. The point of intersection is E.

    e. Click on point E and choose **Annotate** → Coordinates. In the open edit box type (x[2],0).

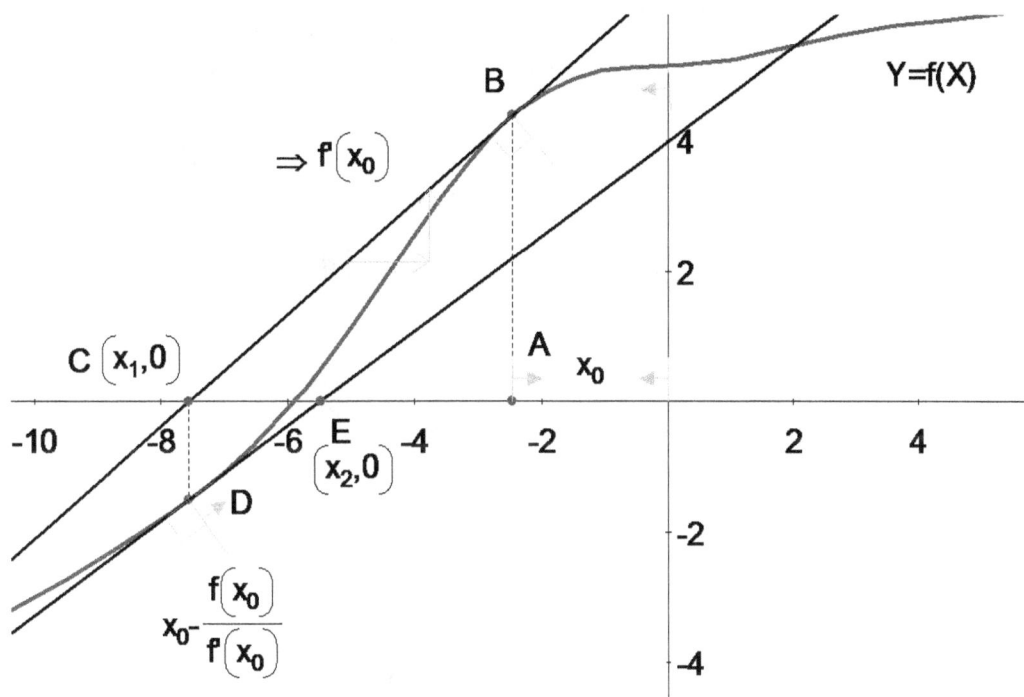

Q5. Which iteration, x_1 or x_2, is closer to the x-intercept of $f(x)$?

A5: The iteration x_2 is closer to the x-intercept of $f(x)$ than x_1.

Note: This answer assumes that students selected their initial estimate in some local proximity of x-intercept. If that is not the case, suggest students to move point A closer to the x-intercept of $f(x)$

Q6. Assume that you completed n iterations and now the point $(x_n, 0)$ is relatively close to the x-intercept of $f(x)$. What is the recursive equation for the $(n + 1)^{th}$ iteration, x_{n+1}?

A6: Since we draw the tangent line to find the next iteration, the equation is

$$y = f(x_n) + f'(x_n)(x_{n+1} - x_n) = 0, \text{ so } x_{n+1} = x_n - \frac{f(x_n)}{f'(x_n)}.$$

8. Use the software to verify your equation.

   a. Delete points D and E. Change the parameter of points A and B to x_n (double-click on x_0 and re-type) and move A closer to the x-intercept.

b. Click point C (the point of intersection between the tangent line and the *x*-axis) and choose **Calculate** → Coordinates. (If you didn't delete it, it will still be there.)

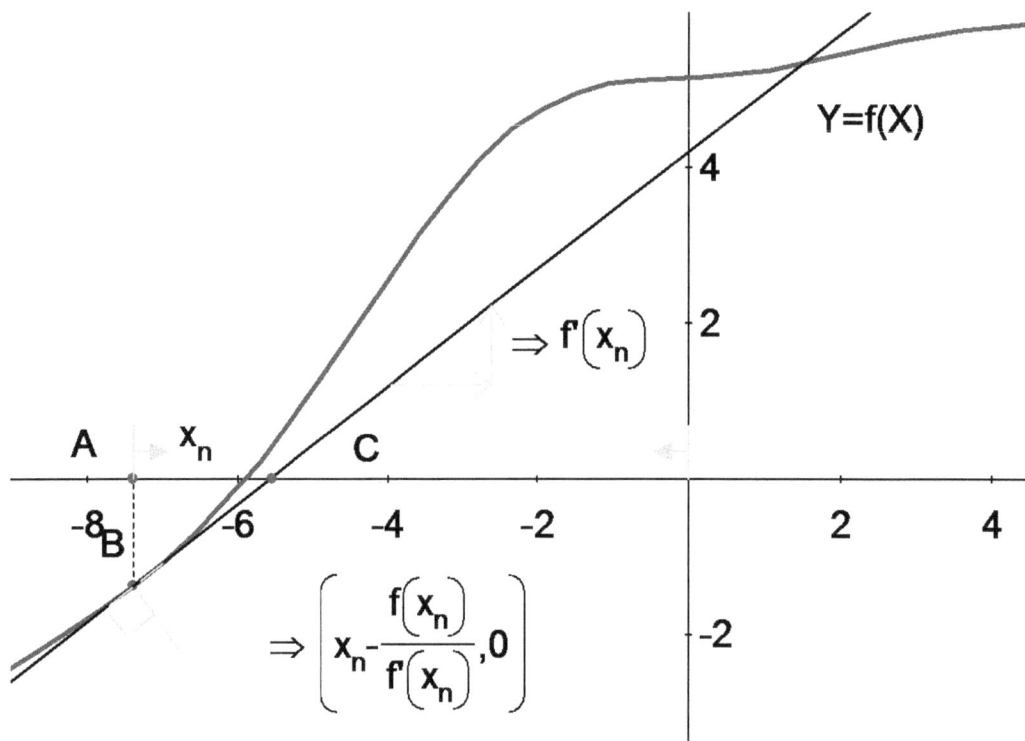

Q7. Which iteration, x_n or x_{n+1}, is closer to the *x*-intercept of *f*(*x*)?

A7: *The iteration x_{n+1} is closer to the x-intercept than the x_n.*

Q8. The precision of the iteration is defined as the absolute difference between the final iteration and the previous iteration. What is the equation for the precision?

A8: $\left| x_{n+1} - x_n \right| = \left| \dfrac{f(x_n)}{f'(x_n)} \right|$

WHEN DOES NEWTON'S METHOD FAIL?

In each example below you are given the function and the seed value. Explore each example with the help of *Geometry Expressions* and explain why Newton's method failed in each case:

1. $f(x) = x^3 - 3x - 1, x_0 = 1$

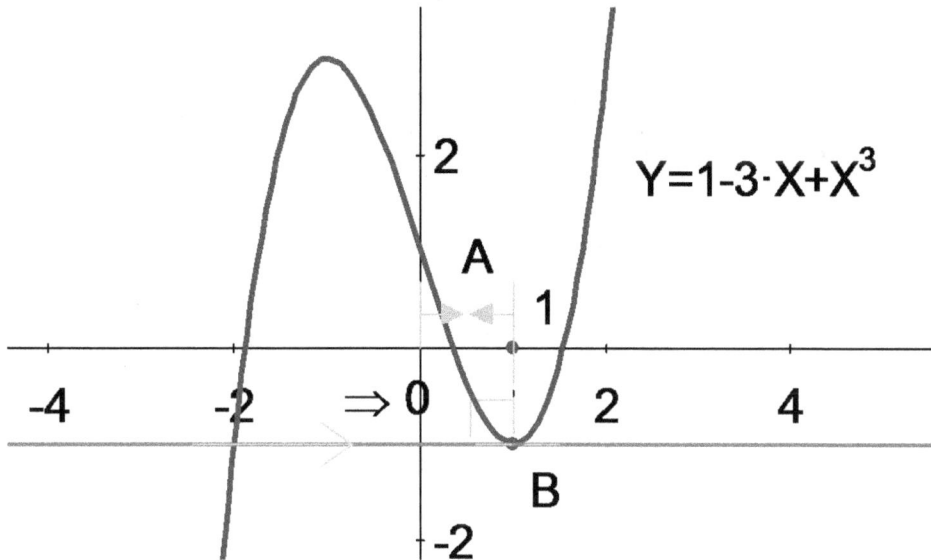

The initial estimate is chosen at the point where the function has a minimum so the tangent line is horizontal and does not have an intersection with the x-axis. Thus the iteration is impossible. Reason: bad choice of initial estimate.

2. $f(x) = x^3 - 2x + 2, \; x_0 = 0$

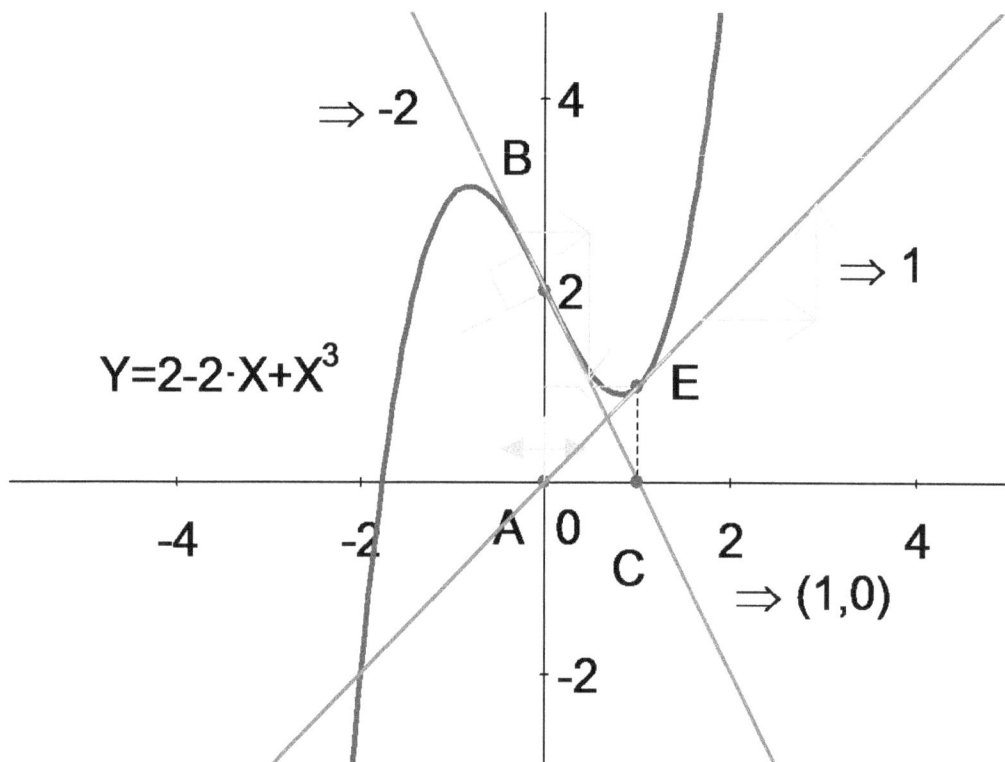

$\Rightarrow -2$

$\Rightarrow 1$

$Y = 2 - 2 \cdot X + X^3$

$\Rightarrow (1,0)$

If you complete the constructions for the first 2 iterations: $x_0 = 0$, $x_1 = 1$, $x_2 = 0$, a cyclic non-convergent process is created. A bad choice of the initial estimate causes oscillations between the guesses back and forth.

3. $f(x) = \arctan(x),\ x_0 = 1.5$

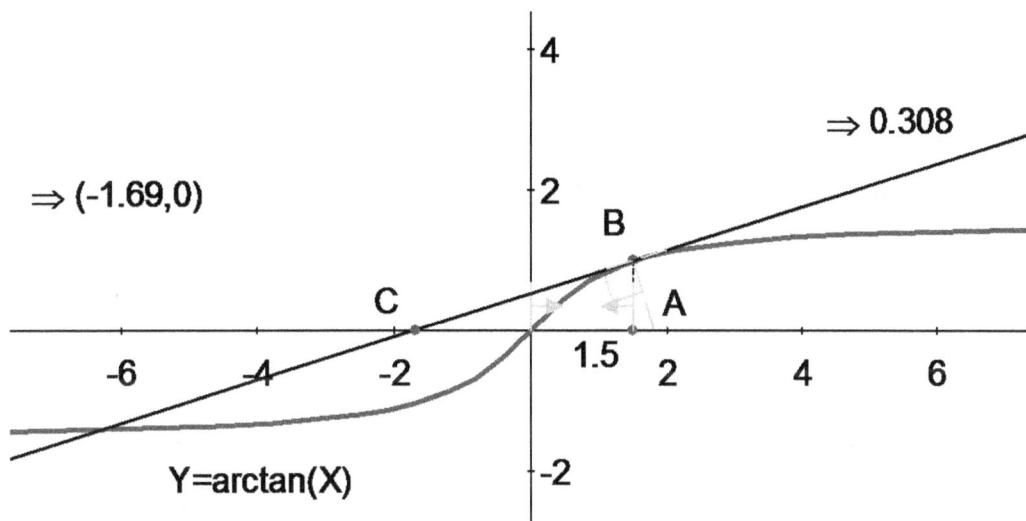

$\Rightarrow 0.308$

$\Rightarrow (-1.69, 0)$

Y=arctan(X)

A bad choice of the initial estimate causes the iterative process to diverge. A more detailed analysis reveals that if $|x_0| > \dfrac{\pi}{2}$ then the iterative process is diverging; if $|x_0| = \dfrac{\pi}{2}$ then the iterative process is cyclic, and only if $|x_0| < \dfrac{\pi}{2}$ is the iterative process converging.

4. $f(x) = |x|^{1/3},\ x_0 = a$, where a is any real number.

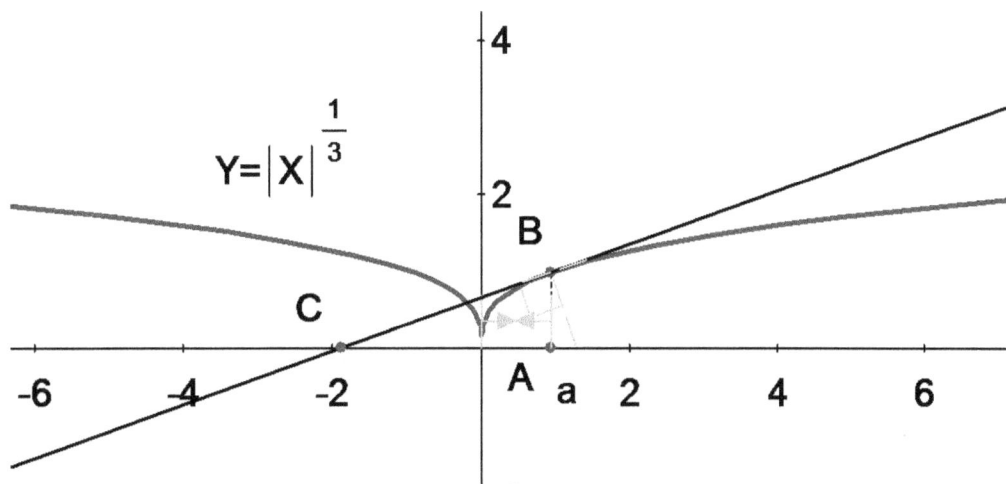

$Y = |X|^{\frac{1}{3}}$

For all x≠0 the derivative of the function is monotonically decreasing so the quotient in the recursive formula is getting larger for each consecutive iteration leading to a diverging process for all initial values except x = 0.

Q10. Based on your observations above, what are the conditions on the function and on the seed value for Newton's method to work?

A10: The function needs to be differentiable on the interval near its zero. The initial estimate should be relatively close to the zero of the function. It works better for functions with low curvature. The method may fail if

1. *the initial value is too far from the true zero*

2. *the derivative of the function is not continuous*

3. *the derivative is zero. Similarly, when the derivative is close to zero, the tangent line is nearly horizontal and hence may "shoot" way past the desired root.*

Newton's Method

Exploration 3.3: In this activity you will explore the Newton's method. Consider a function $y = f(x)$ such that $f(c) = 0$. Start with an initial guess which is reasonably close to the true root and then approximate the function by its tangent line. Compute the x-intercept of this tangent line. This x-intercept will typically be a better approximation to the function's root than the original guess, and the method can be iterated.

EXPLORING NEWTON'S METHOD

1. Draw a function $y = f(x)$. Modify the shape of the function as needed to make sure that the graph of the function crosses the x-axis.

2. Let x_0 be an initial estimate for the x-intercept of the function $f(x)$ (this is also called the *seed* value). Draw this point on the x-axis.

3. Draw a tangent line to the graph of the function at a point $(x_0, f(x_0))$.

Q1. What is the slope of the tangent line at this point?

4. Verify your result using *Geometry Expressions.*

5. Construct an intersection of the tangent line and the x-axis. This point is your 1st iteration. Call it x_1.

Q2. Which is the better approximation for the x-intercept of $f(x)$, x_0 or x_1? Why?

Q3. How did you determine x_1?

6. Use the software to verify your equation.

Q4. Repeat the same process, starting with the 1st iteration, x_1 to find 2nd iteration, x_2. What is x_2 in terms of x_1?

7. Complete this process graphically and construct the point x_2 on the x-axis.

Q5. Which iteration, x_1 or x_2, is closer to the x-intercept of $f(x)$?

Q6. Assume that you completed n iterations and now the point $(x_n, 0)$ is relatively close to the x-intercept of $f(x)$. What is the recursive equation for the $(n + 1)^{st}$ iteration, x_{n+1}?

8. Use the software to verify your equation.

Q7. Which iteration, x_n or x_{n+1}, is closer to the x-intercept of $f(x)$?

Q8. The precision of the iteration is defined as the absolute difference between the final iteration and the previous iteration. What is the equation for the precision?

WHEN DOES NEWTON'S METHOD FAIL?

In each example below you are given the function and the seed value. Explore each example with the help of the software and explain why Newton's method failed in each case:

1. $f(x) = x^3 - 3x - 1$, $x_0 = 1$

2. $f(x) = x^3 - 2x + 2$, $x_0 = 0$

3. $f(x) = \arctan(x)$, $x_0 = 1.5$

4. $f(x) = |x|^{1/3}$, $x_0 = a$, where a is any real number

Q10. Based on your observations above, what are the conditions on the function and on the seed value for Newton's method to work?

3.4 Rectangle in a Semicircle

Exploration 3.4: A rectangle is inscribed in a semicircle of radius r. What is the maximum value of the area of this rectangle?

SUMMARY

Mathematics Objectives:

- Find the rectangle inscribed in a semicircle that has maximum area using both, geometric and calculus approaches.

- Justify the solution by two different methods, with and without calculus.

Vocabulary:

- Optimization

Pre-requisites:

- Pythagorean Theorem

- Product rule and chain rule of differentiation

- 1st and 2nd derivative tests

Problem Notes:

- This is a classic optimization problem of finding the maximum of a function on an open interval. *Gx* allows students to visualize a solution to the problem by dragging the vertices of the rectangle and observing changes in its area.

- Students determine the expression for the area function, plot this function, and explore the existence of a maximum by using the tangent line to the graph of the function. Students can drag the tangent line until they find the coordinates of the point where the derivative of the function is zero and the function is changing from increasing to decreasing. By doing this, they see a visual representation of the 1st and 2nd derivative tests.

- In order to find an exact maximum of the function, students differentiate the function by hand and use the software to confirm their derivative formula, as they calculate the slope of the tangent line.

- Discuss with the students geometric symmetry of the problem and algebraic symmetry of the solution.

Technology skills:

- Draw: circle, line segment, polygon, function

- Constrain: radius, point incident to a segment, perpendicular, length

- Construct: tangent line

- Calculate: area, slope

Extension:

What is the maximum area of a rectangle inscribed in a ellipse with the equation $4x^2 + 9y^2 = 1$.

STEP-BY-STEP INSTRUCTIONS

CONSTRUCTION

1. Draw a circle and constrain its radius to be r.

   a. Use **Toggle grid and axes** to hide the axes and the grid.

   b. Choose **Draw** → Circle. Two points will be displayed: Point A is the center of the circle and Point B is on the circle. Select point B and hide it, using **View** → Hide.

   c. Select the circle and choose **Constrain** → Radius. Type r for the radius of the circle. Lock the value of r in the **Variables** window.

2. Draw a diameter of the circle.

   a. Draw a chord CD on the circle by choosing **Draw** → Segment. Select the chord and the point A and choose **Constrain** → Incident.

3. Construct a rectangle so that two of its vertices are on the diameter and the other two vertices are on the circle.

   a. Choose **Draw** → Polygon and draw a quadrilateral with two vertices on the diameter and two vertices on the circle.

   b. Select side EF and the diameter of the circle, not the side of the quadrilateral, and choose **Constrain** → Perpendicular.

---

c. Select segments EF and FG and choose **Constrain** → Perpendicular. Finally, select FG and GH and choose **Constrain** → Perpendicular.

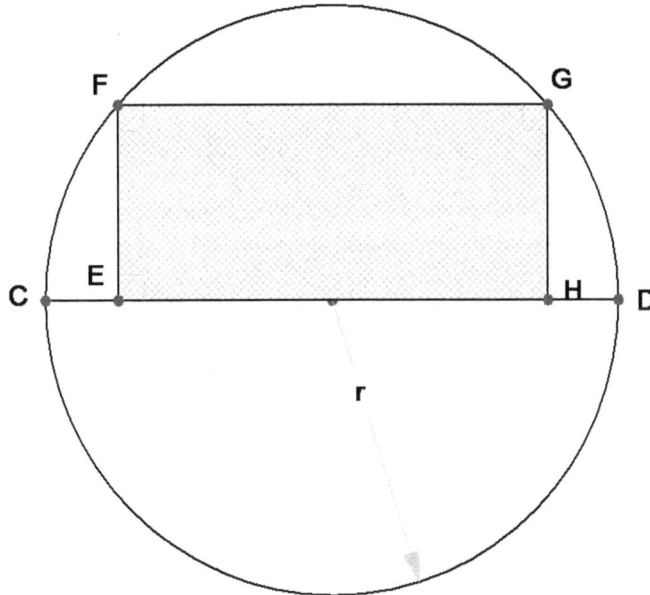

4. Drag one of the vertices of the rectangle and observe any changes in the area of the rectangle.

Q1. Does the area of the rectangle achieve an extreme value? Explain.

A1. As F is dragged towards C or G towards D, the area of the rectangle is approaching zero. When F and G approach each other, the area of the rectangle is also approaching zero. Since area is always positive, it has to achieve a maximum value somewhere between.

INVESTIGATION OF RECTANGLE AREA

5. Constrain the length of the side of the rectangle that is perpendicular to the diameter as x.

a. Select segment EF and choose **Constrain** → Distance/Length. Type x in the edit box.

Q2. Find an expression for the area of the rectangle in terms of x and r.

A2.   Students can use the Pythagorean theorem to find the length of AE to be $\sqrt{r^2-x^2}$. So the area of the rectangle is: $A = 2x\sqrt{r^2-x^2}$

6.   Use the software to calculate the area of the rectangle and verify your answer to Q2.

   a.   Select the rectangle interior and choose **Calculate** → Area.

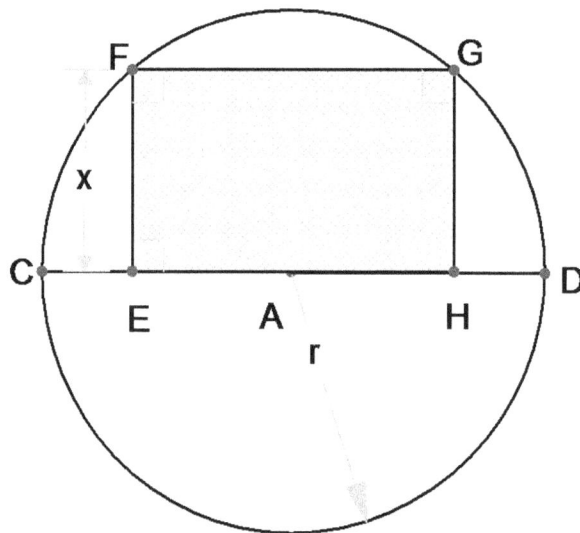

7.   The radius is a constant so the area of the rectangle is a function of *x*. Graph this function.

   a.   Click on the algebraic expression for the area and choose **Edit** → Copy As → String.

   b.   Open a new file. Use **Toggle grid and axes** to display the axes without the grid.

   c.   Select **Draw** → Function. Select Cartesian for the type of the function. Clear the Y= line and select **Edit** → Paste. Press OK.

8. Use the tangent line to confirm that the area has maximum value.

    a. Click the graph of the function and select **Construct** → Tangent.

    b. Drag the tangent line along the curve to observe changes in slope at the location of the maximum.

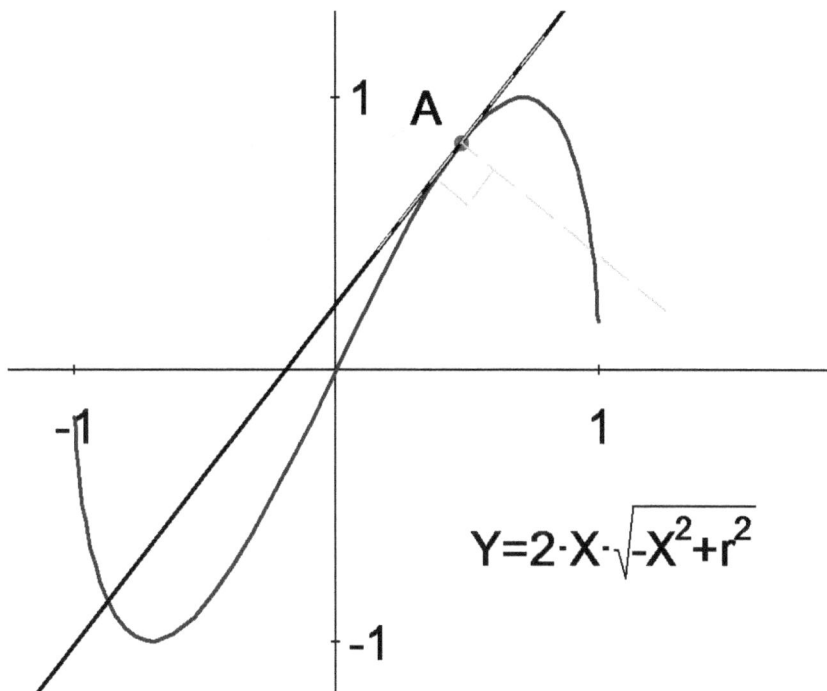

$$Y=2{\cdot}X{\cdot}\sqrt{-X^2+r^2}$$

Q3. How can you determine the exact maximum value of the area of the rectangle?

A3. *Find the coordinates of the point where derivative of the function is zero and the function is changing from increasing to decreasing.*

   a. *Find the derivative using the product rule and the chain rule:*

$$A'(x) = (2x)'\sqrt{r^2 - x^2} + 2x\left(\sqrt{r^2 - x^2}\right)' = \frac{2(r^2 - 2x^2)}{\sqrt{r^2 - x^2}},$$

   *or alternatively, by selecting the tangent line and choosing **Calculate** → Slope.*

$$\Rightarrow \frac{2 \cdot \left(r^2 - 2 \cdot x^2 \right)}{\sqrt{r^2 - x^2}}$$

$$Y = 2 \cdot X \cdot \sqrt{-X^2 + r^2}$$

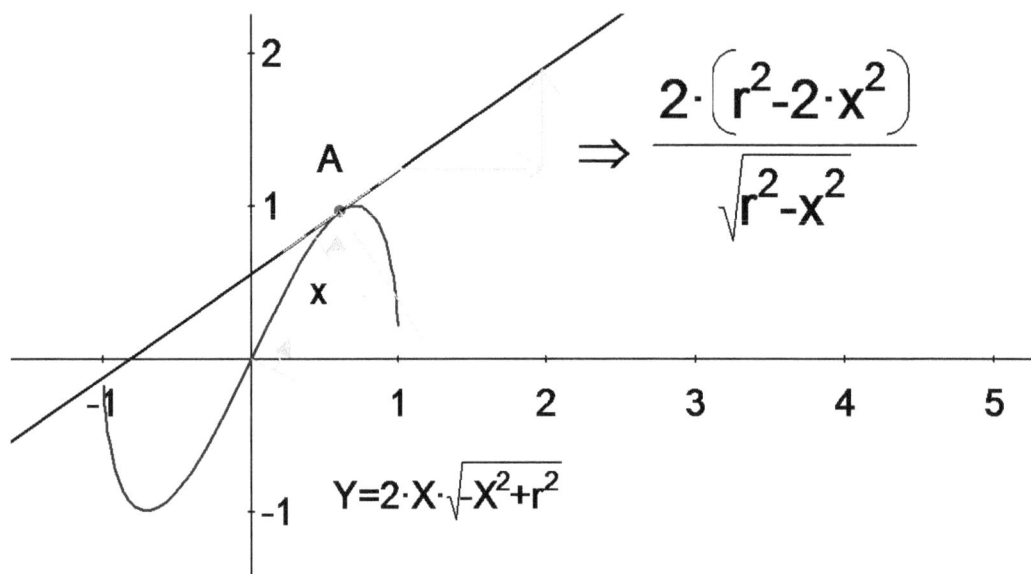

b. Set $r^2 - 2x^2 = 0$, so $x = \dfrac{r\sqrt{2}}{2}$. *The value of the area at this point is* $A = r^2$.

c. *Use 1st or 2nd derivative test to justify that this is a maximum (or refer to the exploration with the tangent line).*

Q4. Did you notice anything special about the rectangle of maximum area? Explain your observations.

A4. *Students may notice that the rectangle of maximum area has the width that is twice the height. For this rectangle AE = FE, and $\angle EAF = \angle GAH = 45°$. Also they could notice that if the problem is expanded to the full circle, the rectangle with the maximum area inscribed in the circle is a square.*

Rectangle in a Semicircle

Exploration 3.4: A rectangle is inscribed in a semicircle of radius r. What is the maximum value of the area of this rectangle?

CONSTRUCTION

1.   Draw a circle and constrain its radius to be r.

2.   Draw a diameter of the circle.

3.   Construct a rectangle so that two of its vertices are on the diameter and the other two vertices are on the circle.

4.   Drag a vertex of the rectangle and observe any changes in the area of the rectangle.

Q1.   Does the area of the rectangle achieve an extreme value? Explain.

INVESTIGATION OF RECTANGLE AREA

5.   Constrain the length of the side of the rectangle that is perpendicular to the diameter as x.

Q2.   Find an expression for the area of the rectangle in terms of x and r.

6.   Use the software to calculate the area of the rectangle and verify your answer to Q2.

7.   The radius is a constant so the area of the rectangle is a function of x. Graph this function.

8.   Use the tangent line to confirm that the area has maximum value.

Q3.   How can you determine the exact maximum value of the area of the rectangle?

Q4.   Did you notice anything special about the rectangle of maximum area? Explain your observations.

3.5 Floating Log

Exploration 3.5: Water flows south in a canal of width a. The canal makes a right angle turn to the east and the width of the canal changes to b as shown in the picture. What is the maximum length L of a log that can float through this turn in the canal?

THIS IS A CLASSICAL OPTIMIZATION PROBLEM. THE MAJOR CHALLENGE FOR STUDENTS IS TO DEVELOP A GEOMETRIC INTERPRETATION OF THE PROBLEM: FIND THE SMALLEST SEGMENT DRAWN THROUGH A GIVEN POINT IN THE 1ST QUADRANT THAT HAS ENDPOINTS ON THE COORDINATE AXES.

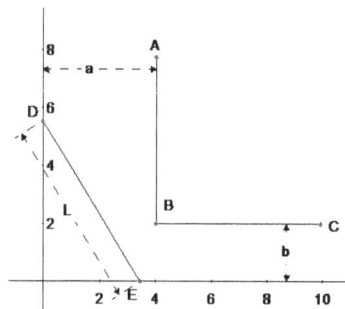

SUMMARY

Mathematics Objectives:

- Determine the maximum length of a segment that can fit around the right-angled corner.

- Justify the solution using calculus

Vocabulary:

- Optimization

Pre-requisites:

- Pythagorean Theorem

- Right triangle trigonometry

- Similar triangle theorems

- Derivatives of trigonometric functions

- Chain rule of differentiation

- 1st and 2nd derivative tests

Problem Notes:

- The major challenge of this problem for students is to develop a geometric interpretation of the problem: finding the segment of maximal length that can fit around the right-angled corner is equivalent to finding the smallest segment drawn through a given point in the 1st quadrant (the corner point) that has endpoints on the coordinate axes. The calculus solution leads to an elegant geometric solution to the problem that may not have been observed without the aid of the software.

- Students first complete the construction of the problem. Then they can vary the length of the segment and drag it around the corner to observe that the segment of maximal length has to go through a point that represents the corner point.

- As soon as they make this observation, they modify their construction to constrain the incidence of the corner point and the segment. Thus, they can now explore the length of the segment as a function of the coordinate of one of the endpoints of the segment. This step provides students with the function to be optimized. Students can determine the function algebraically or with the help of the software.

- It is important to note that the square of the length will have its minimum value at the same point as the length, so it is much simpler to analyze the square of the segment length than the length itself. The minimum of this function can be found with and without calculus. If students do not see the non-calculus solution, use this opportunity to discuss it: The square of the length is the sum of two squares. It reaches minimum value when both squares are equal to zero, so without differentiation students can determine that at the point of the minimum,

$$t = \tan^{-1}\left(-\frac{b}{a}\right).$$

- Students should justify the solution they find using either calculus or geometry considerations.

Technology skills:

- Draw: line segment, function

- Constrain: parallel line, distance, perpendicular, point incident to a line, angle

- Construct: tangent line

- Calculate: area, slope

Extension:

Consider a different canal that has width a. It makes a turn at an angle α remaining the same width. What is the maximum length of a log that can float through this turn in the canal?

STEP-BY-STEP INSTRUCTIONS

CONSTRUCTION

1. Let the x and y axes be the south and west banks of the canal. Draw the east and north banks of the canal using segments parallel to the axes.

   a. Use **Toggle grid and axes** to show the axes without the grid.

   b. Use the hand icon to move the graph to display only the axes and the 1st quadrant.

   c. To represent the canal in Gx, choose **Draw** → Segment and draw a vertical segment AB and a horizontal segment BC so that both segments are in the 1st quadrant.

   d. Select the vertical segment and the y – axis and choose **Constrain** → Parallel.

   e. Select the horizontal segment and the x – axis and choose **Constrain** → Parallel.

2. Constrain the width of the southbound canal to be a, and the width of the eastbound canal to be b. Lock the values of a and b by clicking the lock in the bottom right hand corner of the **Variables** panel.

   a. Select the vertex of the right angle and the y – axis and choose **Constrain** → Distance/Length. In the edit box type a.

   b. Select the vertex of the right angle and x – axis and choose **Constrain** → Distance/Length. In the edit box type b.

   c. Click on each variable, one at a time, in the **Variables** panel. Adjust their values so that they are different (type a new value or use the slider), and then lock these values.

3. Draw a segment representing the log with its endpoints on the x and y axes. Constrain its length to be L.

   a. Choose **Draw** → Segment to draw the log. Place one endpoint on the x-axis and the other endpoint on the y-axis.

b. Constrain the length: Click the segment representing the log and choose **Constrain**
→ Distance/Length. In the edit box type *L*.

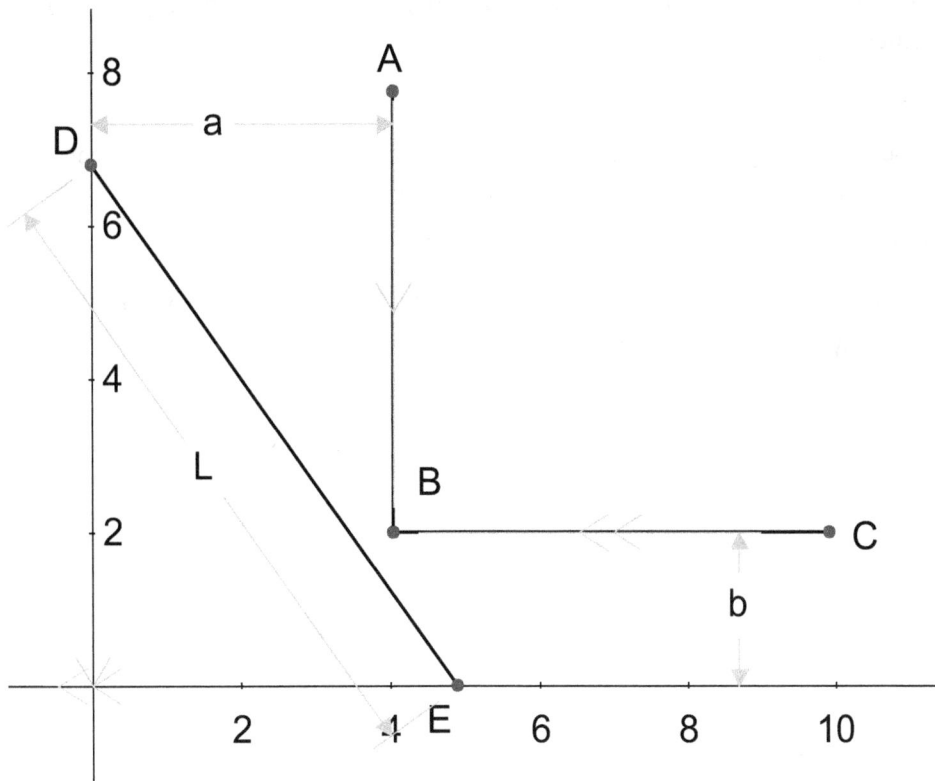

4. Drag the log through the canal. Adjust its length so it can pass through the canal.

Note: In order to observe that the log can or cannot float through the canal, students should drag the segment itself not the endpoints. To adjust the length of the log, students should drag the endpoints of the segment, and then drag the segment again.

Q1. What must be true when the log of the largest possible length can make a turn through the canal?

A1: The longest log that can turn has to touch point B. Thus, we need to compare all the segments through point B with endpoints on the coordinate axes. Then the shortest segment through point B represents the longest log that can go through the canal.

INVESTIGATION OF LOG LENGTH

5. Delete the length constraint from the log and constrain the segment according to your findings in Q1.

    a. Click the constraint L and choose Delete.

    b. Select the segment DE and point B and choose **Constrain** → Incident. This will force the segment to go through point B.

6. Constrain the angle that the segment (the log) forms with the x – axis as t.

    a. Select segment DE and the x – axis. Choose **Constrain** → Angle. In the open edit box type t.

Q2. Find the expression for the length of the segment in terms of a, b, and t.

A2: *Let O be the origin of the coordinate system. From the right triangle DOE we get:*

$$DE = \sqrt{(DO)^2 + (OE)^2} \; ; \qquad DO = b + a\tan(t); \; OE = a + b\cot(t). \qquad Thus,$$

$$DE = \sqrt{(b + a\tan(t))^2 + (a + b\cot(t))^2}$$

7. Calculate the length of the segment using the software and verify your expression.

$$\Rightarrow \sqrt{\left[a+\dfrac{b\cdot\cos(t)}{\sin(t)}\right]^2 + \left[b+\dfrac{a\cdot\sin(t)}{\cos(t)}\right]^2}$$

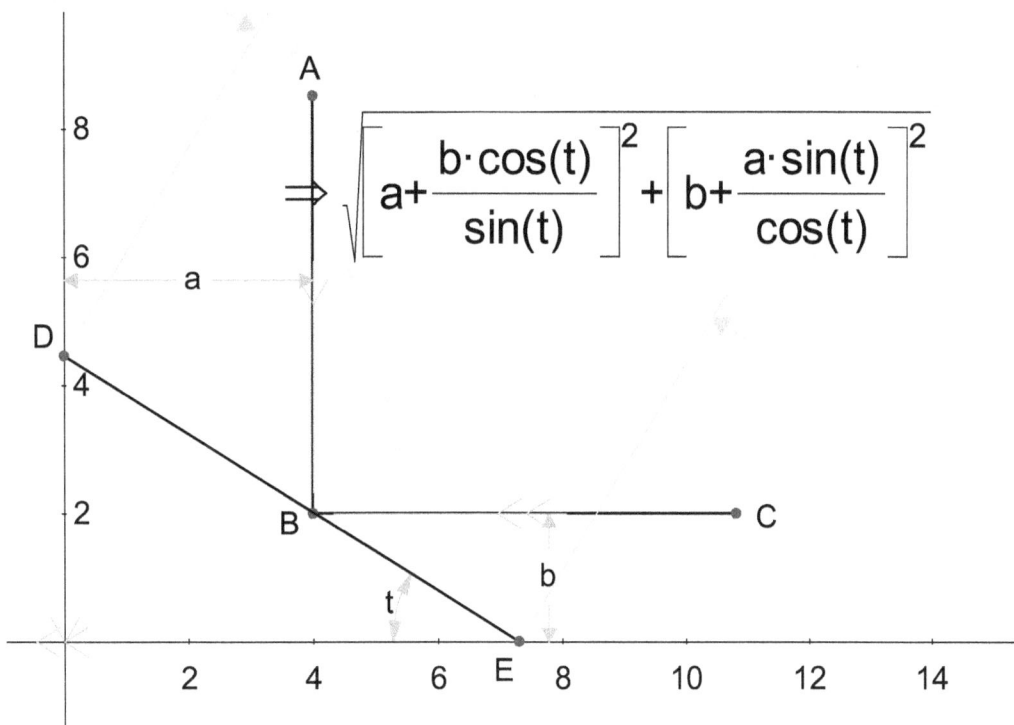

8. Graph the length of the segment as a function of the angle *t*.

    a. Click on the expression for the segment DE and choose **Edit** → Copy As → String.

    b. Open a new document. Use **Toggle grid and axes** to show the axes without the grid.

    c. Move the pane of the graph to see only the axes and 1st quadrant.

    d. Select **Draw** → Function. Select Parametric for the type of the function.

    e. In the X= prompt type T. In the Y= line select **Edit** → Paste.

    f. Choose 0 for Start and 1.6 for End (since the angle can only change from 0 to π/2). Press OK.

9. Use the tangent line to confirm that this function has extreme value.

    a. Select the graph of the function and choose **Construct** → Tangent Line.

    b. Drag the point of tangency along the graph of the function and observe the slope of the behavior of the tangent line.

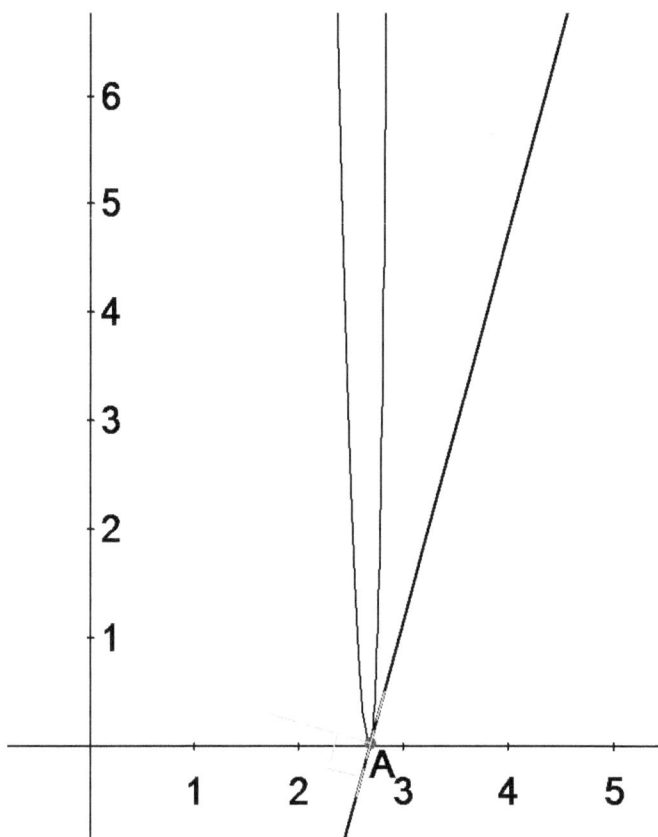

Note: students can graph the square of the length, since it will have an extreme value at the same point as the length. In order to do that, students can delete the exponent ½ to graph the square of the length after completing step 8e. The graph above uses the Y value of (DE)².

Q3. Can you determine the length of the longest log that can float through the canal?

a. Students can take the derivative of $(DE)^2$ by hand:

$$2(a + b\cot(t))\left(-b\csc^2(t)\right) + 2(b + a\tan(t))\left(a\sec^2(t)\right)$$

b. As an alternative, they can also use the software and find slope of the tangent line in symbolic form:

- Select point A and the curve and choose **Constrain** → Point proportional along curve. Type *t* in the edit box.

- Click on the tangent line and choose **Calculate** → Slope.

$$> 2 \cdot \left(a + \frac{a \cdot \sin(t)^2}{\cos(t)^2} \right) \cdot \left(b + \frac{a \cdot \sin(t)}{\cos(t)} \right) + 2 \cdot \left(a + \frac{b \cdot \cos(t)}{\sin(t)} \right) \cdot \left(-b - \frac{b \cdot \cos(t)^2}{\sin(t)^2} \right)$$

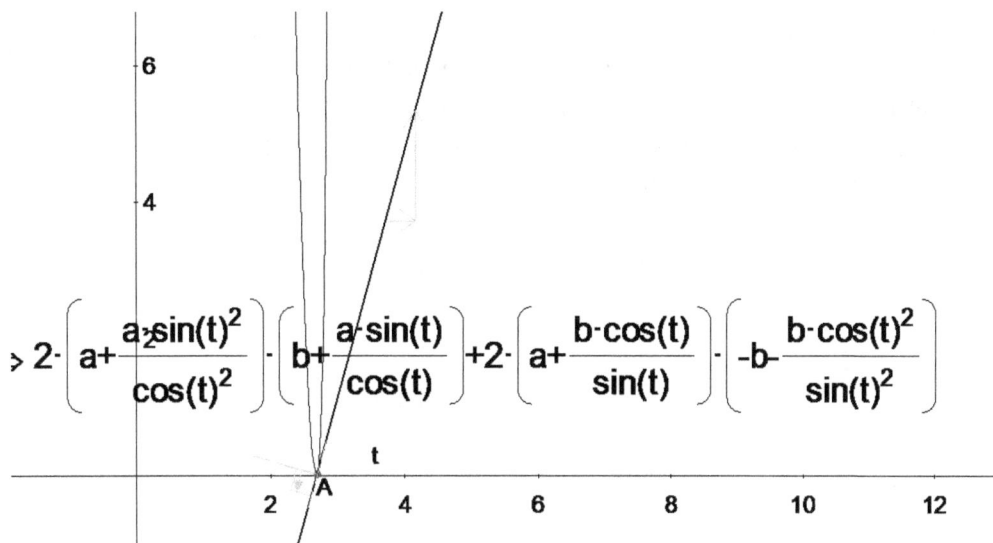

c. Factor this exression and set it equal to zero:

$$2(a + b \cot(t)) \left(-b \csc^2(t) \right) + 2(b + a \tan(t)) \left(a \sec^2(t) \right) =$$

$$2(a \tan(t) + b) \left(-b \csc^2(t) \cdot \cot(t) + a \sec^2(t) \right) = 2(a \tan(t) + b)(a \tan^3 t - b) \cot t \csc^2 t.$$

Since the angle t can only have values in the interval $\left[0, \frac{\pi}{2} \right]$ the only critical

point inside the interval is when $a \tan^3 t - b = 0$. Thus, $t = \tan^{-1} \left(\frac{b}{a} \right)^{1/3}$. The

first derivative test confirms that this is the minimum of the function.

d. Find the length of the log by substitition:

$$L = \sqrt{\left(a + \frac{b}{\tan(t)} \right)^2 + (b + a \tan(t))^2} \Bigg|_{t = \arctan\left(\frac{b}{a}\right)^{1/3}} = \sqrt{\left(a + b \cdot \left(\frac{a}{b} \right)^{1/3} \right)^2 + \left(b + a \cdot \left(\frac{b}{a} \right)^{1/3} \right)^2}$$

$$= \sqrt{\left(a + a^{1/3} b^{2/3} \right)^2 + \left(b + b^{1/3} a^{2/3} \right)^2} = \sqrt{\left(a^{2/3} + b^{2/3} \right)^3}$$

Q4. Verify your solution using the software.

a. Delete all constrains on segment DE.

b. Select DE and choose **Constrain** → Dsitance/Length.

---

c. In the open edit box enter the expression you found for the length using the **Symbols** panel. (If you cannot see the **Symbols** panel, select **View** → Tool Panels → Symbols, the checkbox is a toggle.) You can also type the expression without using the **Symbols** pannel: *SQRT((A^(2/3)+B^(2/3))^3).*

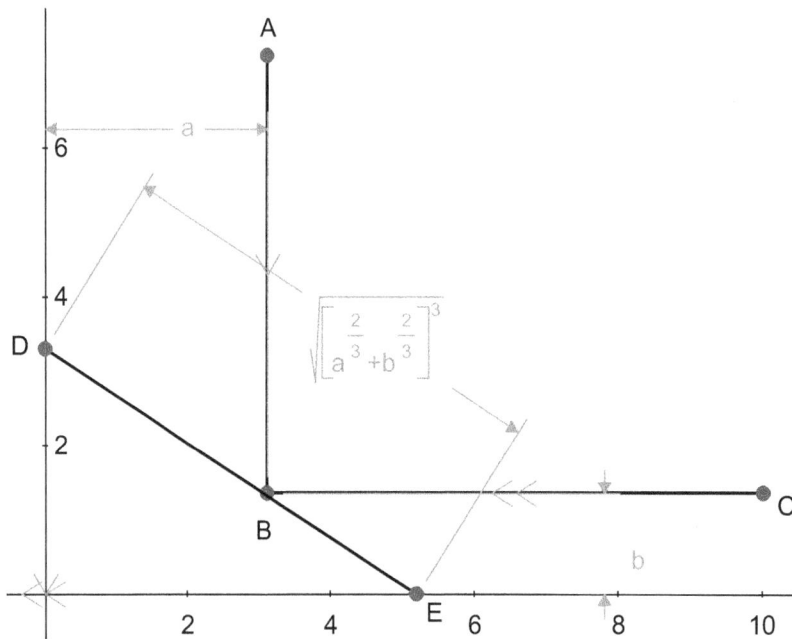

d. Drag the segment to verify that it goes through the corner only touching it at point B.

Floating Log

Exploration 3.5: Water is flowing south in a canal of width *a*. The canal makes a right angle turn to the east and the width of the canal changes to *b* as shown in the picture. What is the maximum length *L* of a log that can float through this turn in the canal?

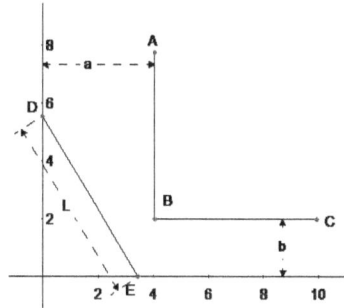

CONSTRUCTION

1. Let the *x* and *y* axes be the south and west banks of the canal correspondingly. Draw the east and north banks of the canal using segments parallel to the axes.

2. Constrain the width of the southbound canal to be *a*, and the width of the eastbound canal to be *b*. Lock the values of *a* and *b* by clicking the lock in the bottom right hand corner of the **Variables** window.

3. Draw a segment representing the log with its endpoints on the *x* and *y* axes. Constrain its length to be *L*.

4. Drag the log through the canal. Adjust its length so it can pass through the canal.

Q1. What must be true when the log of the largest possible length can make a turn through the canal?

INVESTIGATION OF LOG LENGTH

5. Delete the length constraint from the log and constrain the segment according to your observations in Q1.

6. Constrain the angle that the segment (the log) forms with the *x* – axis as *t*.

Q2. Find the expression for the length of the segment in terms of *a*, *b*, and *t*.

7. Calculate the length of the segment using Gx and verify your expression.

8. Graph the length of the segment as a function of the angle, *t*. (*Note: you can graph the square of the length since it will have its extreme value at the same point as the length*).

9. Use the tangent line to confirm that this function has an extreme value.

Q3. Determine the length of the longest log that can float through the canal.

10. Verify your solution using Gx.

3.6 Art Gallery

Exploration 3.6: A painting hangs from a wall. Given the heights of the top and bottom of the painting above the viewer's eye level, how far from the wall should the viewer stand in order to maximize the angle subtended by the painting and whose vertex is at the viewer's eye?

SUMMARY

Mathematics Objectives:

- Determine the distance from the painting that provides maximum viewing angle.

- Justify the solution using calculus

Vocabulary:

- Optimization

Pre-requisites:

- Inverse trigonometric relationships

- Derivatives of trigonometric functions

- Product, quotient, and chain rules of differentiation

- 1st derivative test

Problem Notes:

- The Art Gallery Problem is also known as the problem of Regiomontanus. In 1471, Johann Muller, alias Regiomontanus, posed the question: At what horizontal distance from an elevated rod would a person have to stand such that the appearance of the rod should be a maximum? It has been claimed that this was the first optimization problem in the history of mathematics since antiquity.

- This problem appears in many contemporary calculus books and it involves taking the derivative of an inverse trigonometric function. However, since the viewing angle is always positive and acute, it is not necessary. Students can analyze the tangent of the viewing angle that reaches the maximum value at the same point as the angle itself.

- Students first complete the construction of the problem. Then they can vary the position of the observer to see that the viewing angle does have a maximum value. Students can determine the function algebraically; however, it is time consuming. It is recommended to use the software to obtain an algebraic expression for the viewing angle.

- Students can plot the function of the angle and explore the slope of the tangent line in order to find the maximum viewing angle. The software will produce the expression for the slope of the tangent line, so students will only have to determine the zero of the slope function and justify that the found point is a maximum.

- If the teacher prefers students to do differentiation by hand, it is sufficient to differentiate the tangent of the angle in order to find the maximum angle.

- Non-calculus solutions can be easily developed with the help of *Gx*. The algebraic solution involves the fact that the arithmetic mean of two numbers is always greater than or equal to the geometric mean. A geometric approach uses inscribed angles.

Technology skills:

- Draw: line segment, function,

- Constrain: distance, perpendicular, point proportional along the curve, angle

- Construct: tangent line

- Calculate: angle, slope, coordinates of a point

Extension:

A balloon is rising vertically at a speed v. The height of the balloon (its vertical size) equals h. In the initial moment its lowest point is at height h above the eye level of the observer. At each moment of time the observer moves to a position such that the viewing angle of the balloon is at a maximum. What is the maximum speed of the observer?

STEP-BY-STEP INSTRUCTIONS

CONSTRUCTION

1. Use **Toggle grid and axes** to show the axes without the grid. (We assume that the observer's eye level is represented by the x-axis).

Teacher Notes

2. Draw a segment AB on the *y*-axis to represent the painting. Set the distance from the *x*-axis to point A to be equal to *a* and distance from the *x*-axis to point B to be equal to *b*.

    a. Choose **Draw** → Segment and plot segment AB on the positive *y*-axis.

    b. Select point A and the *y*-axis, and choose **Constrain** → Point proportional. Type *a* in the open edit box. Lock the value of *a* in the **Variables** tool panel.

    c. Select point B and the *y*-axis, and choose **Constrain** → Point proportional. Type *b* in the open edit box. Lock the value of *b* in the **Variables** tool panel.

3. Draw a viewing angle of segment AB from point C on the *x*-axis.

    a. Choose **Draw** → Line Segment and plot segment AC, where point C is on the positive *x*-axis. Then plot segment BC.

    b. Select point C and the *x*-axis, and choose **Constrain** → Point proportional. Type *x* in the open edit box.

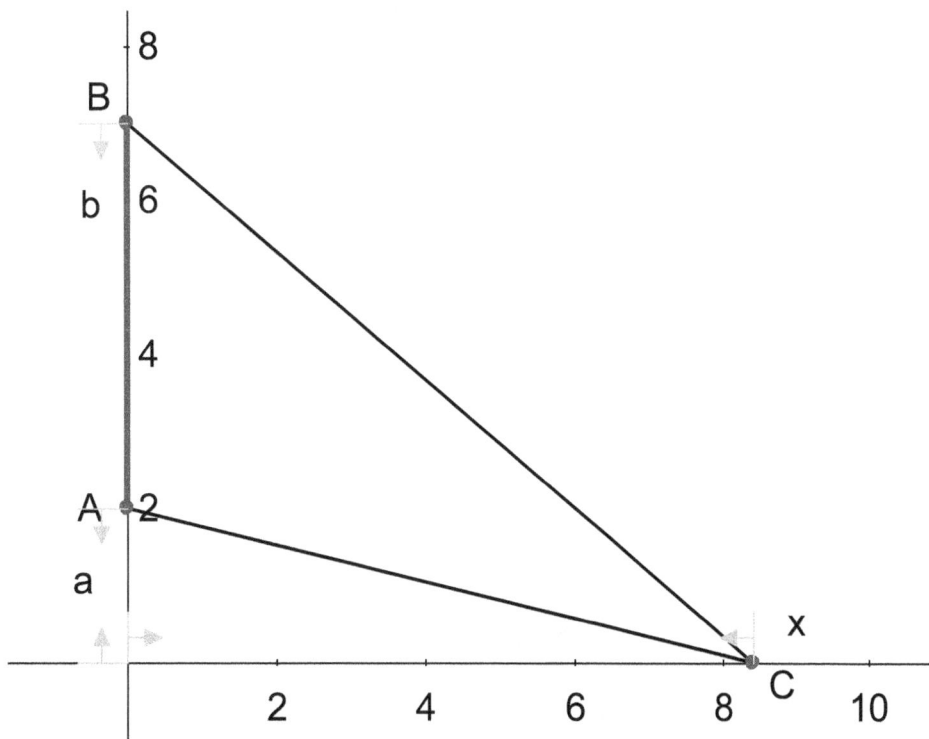

Q1. As you move along the x-axis, does the viewing angle, ∠ACB, change? If so, describe the behavior of ∠ACB. Do you think there is a maximum value for ∠ACB?

A1: *Dragging point C along the x-axis demonstrates that the angle is very small when x is close to the wall, and the angle also gets small when x goes to infinity. Thus, the maximum should exist somewhere in between.*

Note: *To see the viewing angle numerically: select segments AC and BC and choose* **Calculate → Real →** *Angle*

Q2. On what interval is $\angle ACB$ defined? What are the possible values of $\angle ACB$ on this interval?

A2: *$\angle ACB$ is defined on $[0, \infty]$. The range of values of $\angle ACB$ is $[0, \alpha]$, where $\alpha < \dfrac{\pi}{2}$.*

ANALYSIS

4.   Find the viewing angle as a function of x with the help of the software.

    a.   Select segments AC and BC and choose **Calculate → Symbolic →** Angle.

Note: if students calculated numeric values of the angle, they will need to delete the numeric calculations of the angle in order to be able to calculate the symbolic expression. The expression may vary based on the way in which students constrained values a and b. For example, one possible answer could be $\arctan\left(\dfrac{x \cdot (-a + b)}{a \cdot b + x^2}\right)$ *, which is equivalent to the expression shown below.*

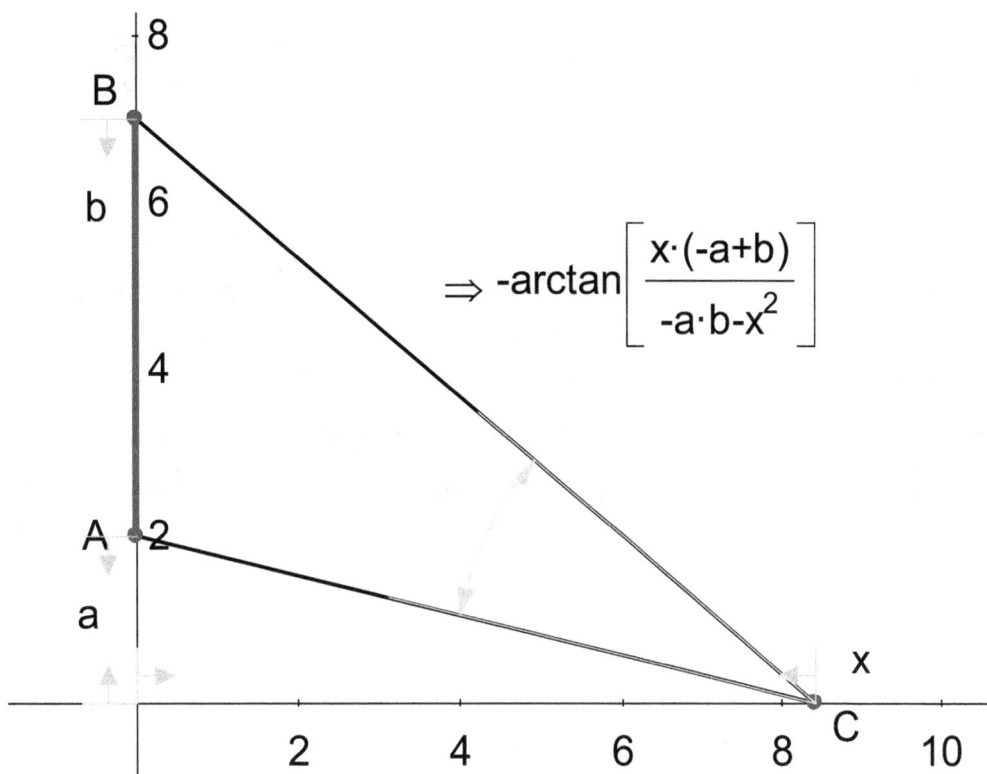

$$\Rightarrow -\arctan\left[\frac{x\cdot(-a+b)}{-a\cdot b-x^2}\right]$$

Q3. Find the maximum of this function.

A3: *When you drag point C it is easy to see that the angle is acute so the tangent of an angle is a positive function of x. Since the tangent of an acute angle has its maximum at the same point as the angle, we will consider* $f(x) = \tan(\angle ACB)$ *for the optimization:*

$$f(x) = \tan\left(-\arctan\left(\frac{x(-a+b)}{-ab-x^2}\right)\right) = -\frac{x(-a+b)}{-ab-x^2} = \frac{x(b-a)}{x^2+ab}.$$

Take the derivative of the function and set it equal to 0:

$$\frac{d}{dx}\left(\frac{(b-a)x}{x^2+ab}\right) = (b-a)\frac{x^2+ab-2x^2}{(x^2+ab)^2} = (b-a)\frac{ab-x^2}{(x^2+ab)^2} = 0, \text{ thus } x = \sqrt{ab}.$$

If $x < \sqrt{ab}$, the derivative is positive, and if $x > \sqrt{ab}$, the derivative is negative, so by the 1st derivative test $x = \sqrt{ab}$ is a maximum point. At this point the angle is equal to $\arctan\left(\dfrac{(b-a)}{2\sqrt{ab}}\right)$.

5.   Plot the function for the angle and find the derivative of the function.

   a.   Select the expression or the function for the viewing angle and choose **Edit** → Copy As → String. Choose **Draw** → Function, select Cartesian for the type of function, and paste the expression for the derivative into the Y= prompt.

   b.   Select the graph of the function and choose **Construct** → Tangent to the curve.

   c.   Select the point of tangency (point D) and the curve, and choose **Constrain** → Point proportional. Type x in the open edit box.

   d.   Select the tangent line and choose **Calculate** → Slope.

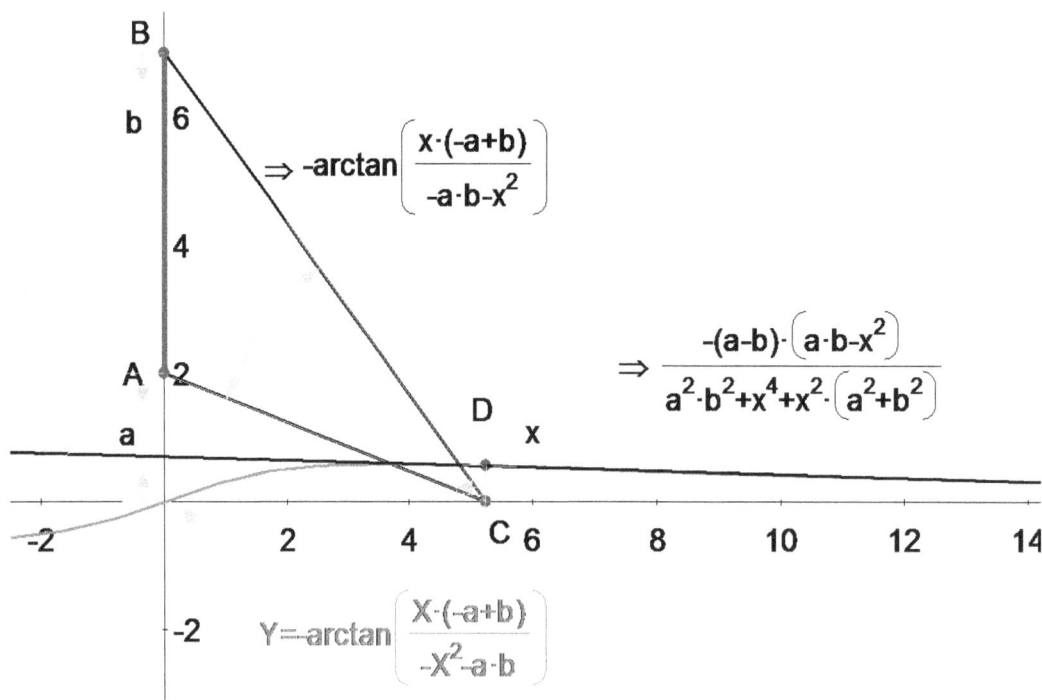

$$\Rightarrow -\arctan\left(\frac{x\cdot(-a+b)}{-a\cdot b-x^2}\right)$$

$$\Rightarrow \frac{-(a-b)\cdot\left(a\cdot b-x^2\right)}{a^2\cdot b^2+x^4+x^2\cdot\left(a^2+b^2\right)}$$

$$Y=\arctan\left(\frac{x\cdot(-a+b)}{-x^2-a\cdot b}\right)$$

Q4. Compare the derivative you calculated with the derivative provided by the software. Did the software confirm your analysis?

A4: Students can drag point D and observe changes in the sign of the slope of the tangent line. The formula for the derivative given by the software provides the same maximum point.

6.   Confirm that the derivative takes a value of *zero* at the maximum point of the function.

   a.   Double-click on the *x*-coordinate of point D and type the coordinate of the maximum: $\sqrt{a \cdot b}$.

   b.   Select point D and choose **Calculate** → Coordinates.

   c.   Select the tangent line to the graph of the function and choose **Calculate** → Slope.

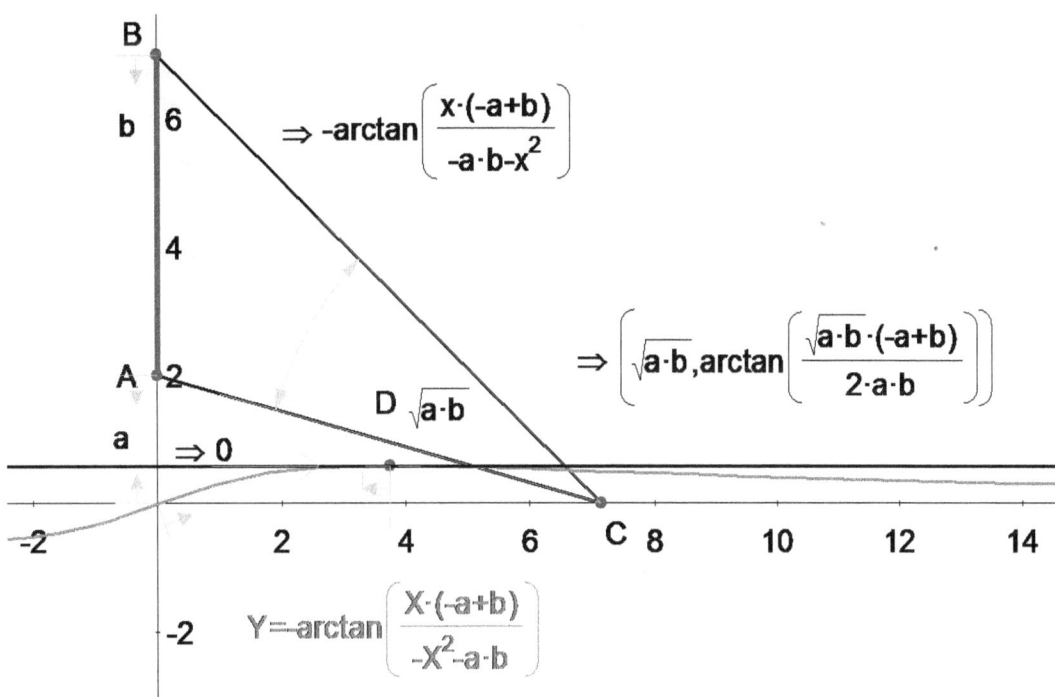

$$\Rightarrow -\arctan\left(\frac{x \cdot (-a+b)}{-a \cdot b - x^2}\right)$$

$$\Rightarrow \left(\sqrt{a \cdot b}, \arctan\left(\frac{\sqrt{a \cdot b} \cdot (-a+b)}{2 \cdot a \cdot b}\right)\right)$$

$$Y = \arctan\left(\frac{X \cdot (-a+b)}{-X^2 - a \cdot b}\right)$$

Q5.   What are the coordinates of the maximum point and what is the value of the slope of the tangent line at this point?

A5: The slope of the tangent line is zero, confirming that the derivative is zero. The maximum value of the angle provided by the software is the same as found by optimization.

Art Gallery

<u>Exploration 3.6:</u> A painting hangs from a wall. Given the heights of the top and bottom of the painting above the viewer's eye level, how far from the wall should the viewer stand in order to maximize the angle subtended by the painting and whose vertex is at the viewer's eye?

CONSTRUCTION

1. Use **Toggle grid and axes** to show the axes without the grid. (We assume that the eye level is represented by the x-axis).

2. Draw a segment AB on the y-axis that represents the painting. Set the distance from the x-axis to point A to be equal to a and the distance from the x-axis to point B to be equal to b.

3. Draw a viewing angle of the segment AB from point C on the x-axis.

Q1. As you move along the x-axis, does the viewing angle $\angle ACB$ change? If so, describe the behavior of $\angle ACB$. Do you think there is a maximum value for $\angle ACB$?

Q2. On what interval is $\angle ACB$ defined? What are the possible values of $\angle ACB$ on this interval?

ANALYSIS

4. Find the viewing angle as a function of x with the help of the software.

Q3. Find the maximum of this function.

5. Plot the function of the angle and find the derivative of the function.

Q4. Compare the derivative you calculated with derivative provided by the software. Did the software confirm your analysis?

6. Confirm that the derivative has a zero value at the maximum point of the function.

Q5. What are the coordinates at the maximum and what is the value of the slope of the tangent line at this point?

4. Integrals and Their Applications

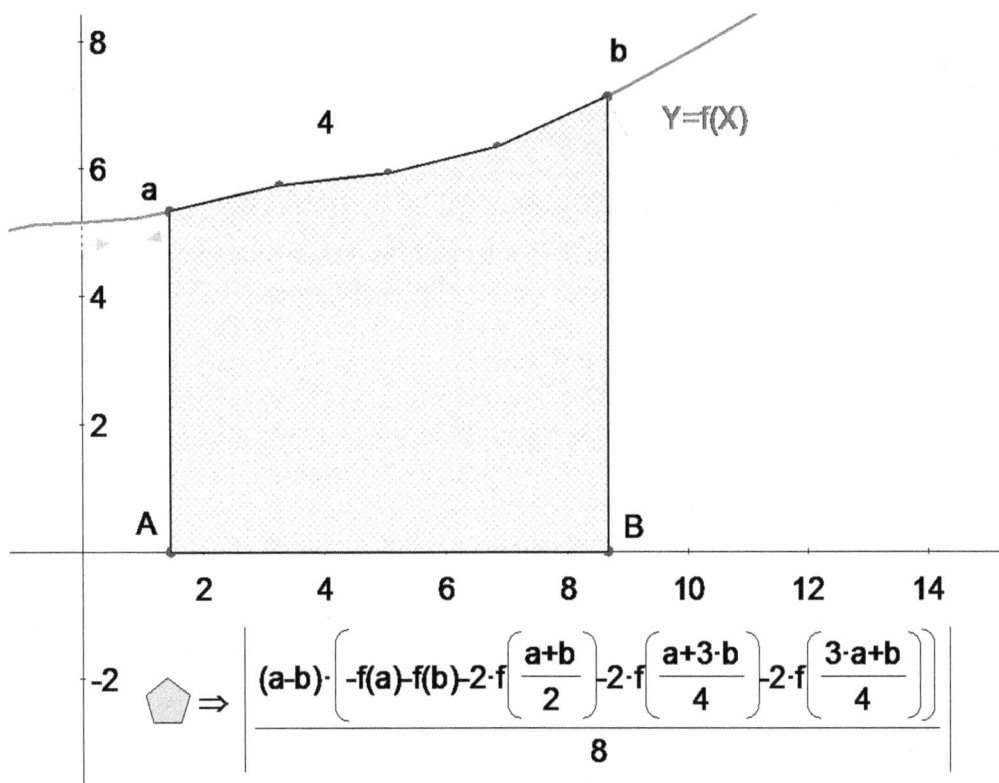

$$\boxed{\Rightarrow} \quad \left| \frac{(a-b) \cdot \left(-f(a) - f(b) - 2 \cdot f\left(\frac{a+b}{2} \right) - 2 \cdot f\left(\frac{a+3 \cdot b}{4} \right) - 2 \cdot f\left(\frac{3 \cdot a+b}{4} \right) \right)}{8} \right|$$

4.1 Representation of the Antiderivative

Exploration 4.1: Suppose you were asked to find a function F whose derivative is $f(x)=2x$. From your knowledge of derivatives, you would probably say that $F(x)=x^2$, since $\frac{d(x^2)}{dx}=2x$. Is this the only function with the given derivative? How many functions can you find with this derivative? What is the relative position of their graphs? You will explore these questions starting with a specific example and then generalizing for the function $y = f(x)$.

SUMMARY

Mathematics Objectives:

- Determine that there are infinitely many antiderivatives of a given function.

- Determine that any two functions can be antiderivatives of a given function if and only if they differ from each other by a constant. Their graphs can be examined by a vertical translation of one to another.

Vocabulary:

- Differential equation

- Particular antiderivative

- General antiderivative (also called the family of antiderivatives)

- Vector of translation

- Constant of integration

Pre-requisites:

- Derivative as a slope of a tangent line.

- Translation

- Basic rules for differentiation

Problem Notes:

- In this exploration students develop an understanding of an important calculus theorem about representation of antiderivatives: $F(x)$ and $G(x)$ are antiderivatives of $f(x)$ on an interval if and only if $G(x) = F(x) + C$ for all points on this interval, where C is a constant.

- Students first explore the antiderivative of a specific function, $f(x) = 3x^2$. They determine a particular and general antiderivative, plot the graphs of the particular and general antiderivative, and explore the meaning of the constant of integration, both algebraically and graphically.

- One of the powerful features of *Geometry Expressions* is the ability to work with generic functions. Students move from the initial specific function to a generic function, $f(x) = \dfrac{dF}{dx}$, and generalize their findings about the representation of antiderivatives and the meaning of the constant of integration, both algebraically and graphically.

- Some answers in this exploration use the slang "parallel" for the graphs of functions. Although this is not a proper mathematics term, it has a strong intuitive meaning for the students, and that is the reason it is used as an example of a possible student answer. When we say he graphs of functions are "parallel", we mean that for points on the function graphs with the same value of x the tangent lines to those graphs are parallel. "Parallel" function graphs are identified by a vertical translation from one to another.

Technology skills:

- Draw: function

- Constrain: point proportional along a curve

- Calculate: slope

- Construct: tangent line, translation, trace

Extensions:

1. Find all functions for which the second derivative is $f(x) = x$.

2. Find all functions that pass through the point (1, 1) for which the second derivative is $f(x) = x$.

STEP-BY-STEP INSTRUCTIONS

THE ANTIDERIVATIVE OF $f(x) = 3x^2$

1. Draw a function $f(x) = 3x^2$.

    a. Use **Toggle grid and axes** to show the axes without a grid.

    b. Choose **Draw** → Function. Select Type → Cartesian. In the Y = prompt type 3*x^2.

Q1. If $f(x) = \dfrac{dF}{dx}$, then $F(x)$ is called the antiderivative of $f(x)$. What is $F(x)$ for this function?

A1: $F(x) = x^3$, since $(x^3)' = 3x^2$. Students may give various answers that differ by a constant from this expression.

2. Draw an antiderivative $F(x)$ on the same graph as $f(x)$.

    • Choose **Draw** → Function. Select Type → Cartesian. In the Y = prompt type x^3.

3. Verify your formula graphically, using a tangent line.

    a. Select the plot of the function $F(x)$ and choose **Construct** → Tangent. The tangent line and a point of tangency will be displayed (point A).

    b. Select point A and the graph of F and choose **Constrain** → Point proportional. Type x in the edit box to constrain the x-coordinate of point A to be x.

    c. Select the tangent line and choose **Calculate** → Slope.

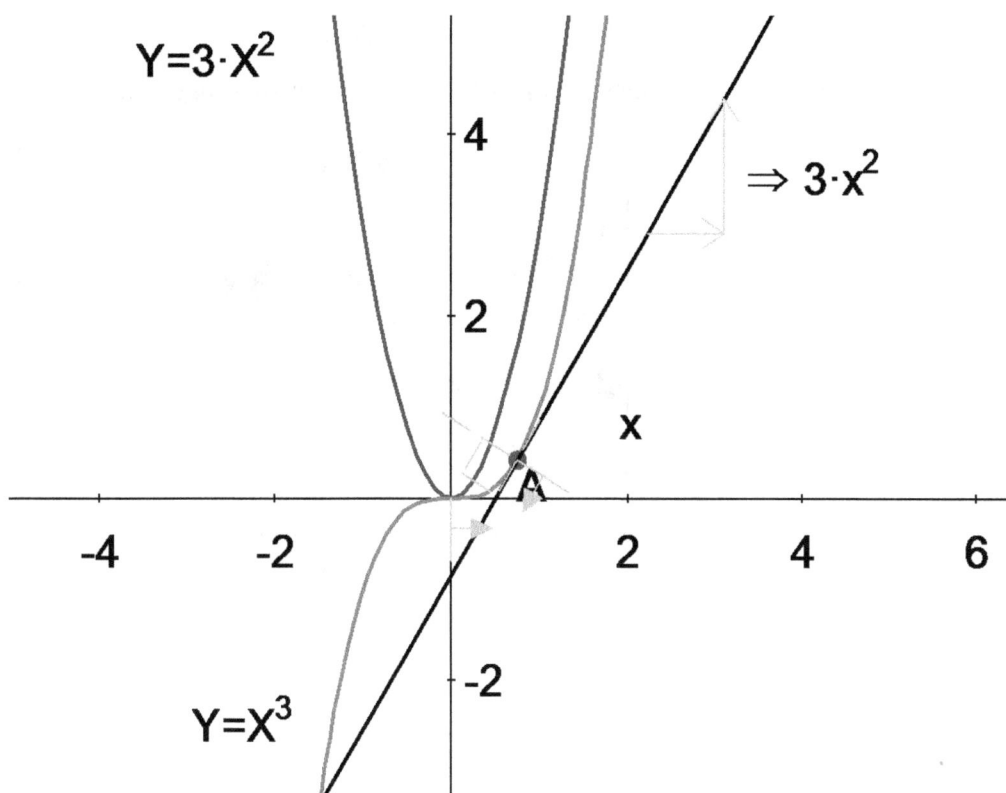

Q2. Can you think of other functions that have the same derivative?

A2: Any function that is different from $F(x) = x^3$ by a constant, for example $G(x) = x^3 + 1$.

Q3. How can you represent the entire family of these functions by a formula?

A3: $G(x) = x^3 + C$, where C = constant.

4. Plot the antiderivative in its general form. Verify that this antiderivative has the same derivative as your $F(x)$.

   a. Choose **Draw** → Function. Select Type → Cartesian. In the Y = prompt type X^3+C.

   b. Select the plot of the function $G(x)$ and choose **Construct** → Tangent. The tangent line and point of tangency will be displayed (point B).

   c. Select point B and the graph of G and choose **Constrain** → Point proportional. Type x in the edit box to constrain the x-coordinate of point B to be x.

d. Select the tangent line and choose **Calculate** → Slope.

Note: you may want to delete or hide the graph of f(x) to unclutter your diagram.

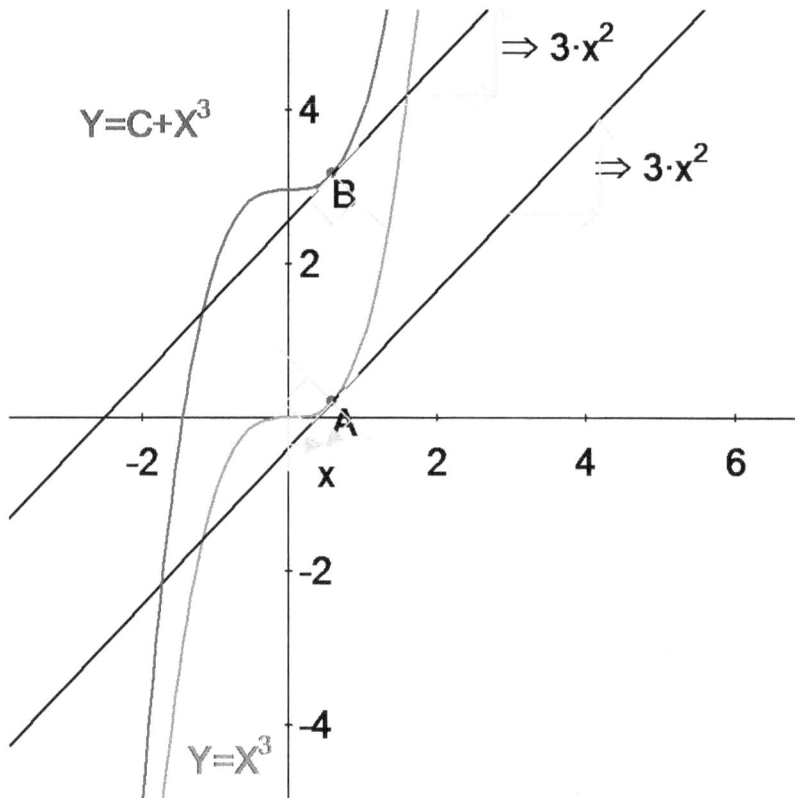

Q4. How can you determine any antiderivative geometrically from the given antiderivative?

A4: All functions G(x) can be determined from the function F(x) as a translation along the y-axis by a vector (0, C)

5. Verify your conjecture with the software

a. Click on the graph of the function $F(x) = 3x^2$ and choose **Construct** → Translation.

b. Click on the origin and then on an arbitrary point on the y-axis to construct the vector of translation $\overrightarrow{CD}$. The image of $F(x)$ will appear on the graph.

c. Click on the point D and y-axis, and choose **Constrain** → Point proportional along curve. In the open edit box type *C* (it's case sensitive!).

d. The graph of the image will coincide with the graph of $G(x)$.

e. Drag point D along the y-axis and observe that the translated image of $F(x)$ by vector $\overrightarrow{CD} = (0, C)$ remains on top of $G(x)$. $G(x) = F(x) + C$.

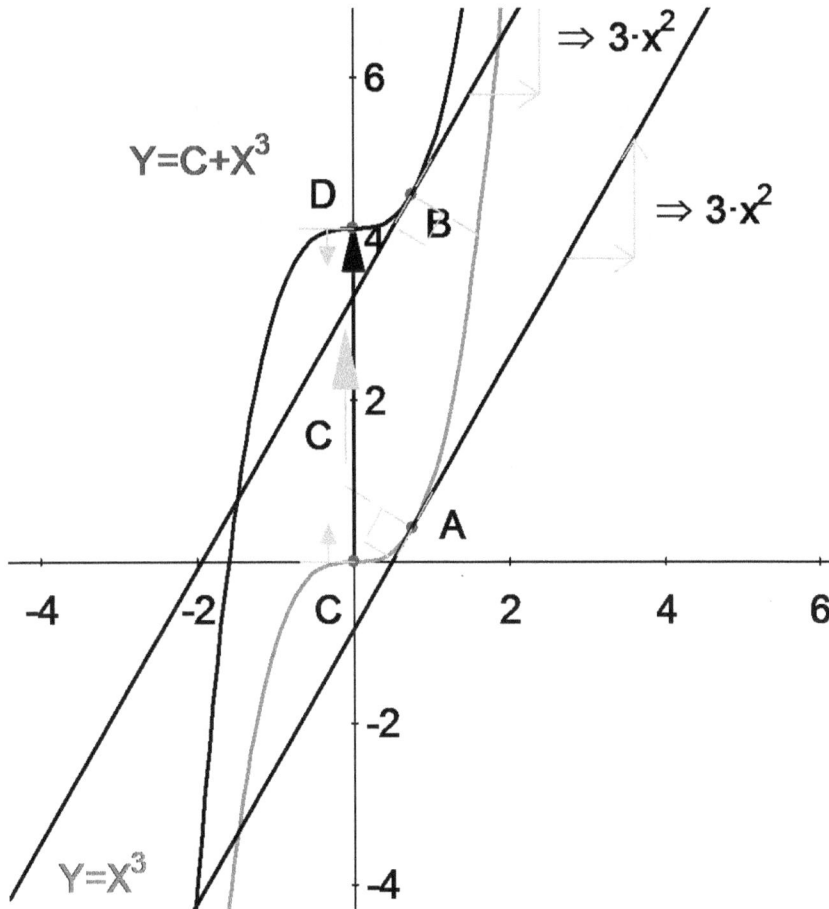

*Note: students can use **View** → Toggle Hidden to toggle between the two curves that are on top of each other. It will be easier for them to see that these are two different plots, if they change the color of one of the curves.*

ANTIDERIVAIVE OF $f(x) = \dfrac{dF}{dx}$

6. Replace $y = x^3$ with $y = F(x)$, and $y = x^3 + C$ with $y = F(x) + C$

    a. Double-click the expression for the function $y = x^3$ and type $F(x)$.

b.   Double-click the expression for the function $y = x^3 + C$ and type $F(x) + C$.

c.   Adjust the shape of $F(x)$ as needed, by dragging the curve in 5 different places.

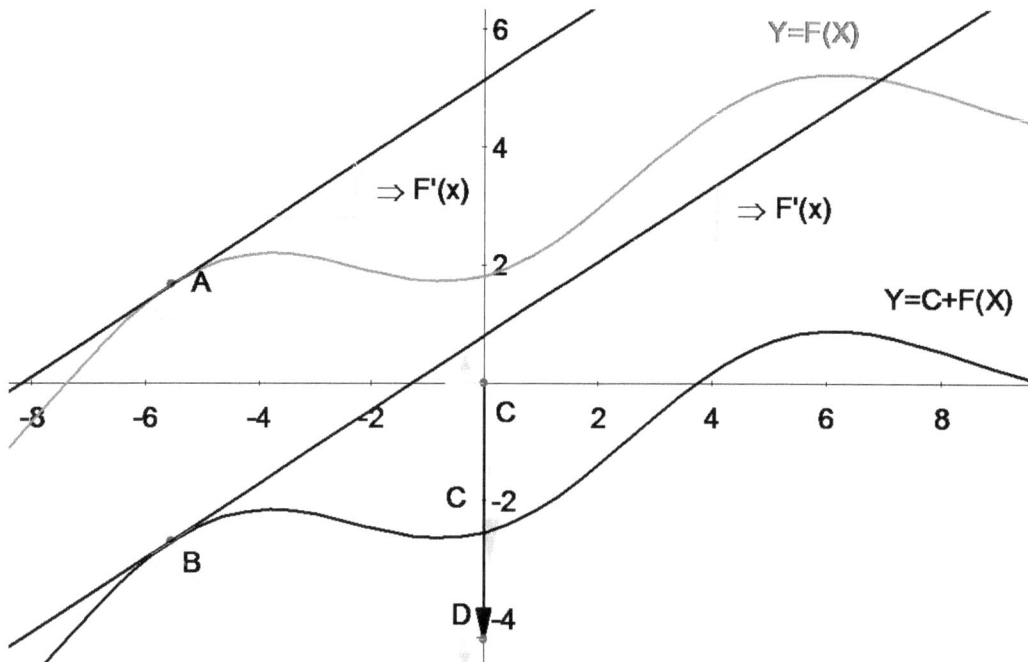

Q5.   Compare the graph of the general antiderivative $F(x) + C$ with the translated image of $F(x)$ by vector $\overrightarrow{CD} = (0, C)$. Does your answer to Q4 hold for a generic function $F(x)$?

*A5: Yes. For any differentiable function F(x) the general antiderivate F(x) + C can be determined as a translation of the graph of F(x) by the vector (0, C). You can drag point D of the vector CD or change the value of C using the slider in the **Variables** panel. The plot of the translated image always stays on top of the function F(x) + C.*

*Note: Use **View** → Toggle Hidden to toggle between the two curves.*

Q6.   What is the slope of the tangent line at point $A(x, F(x))$? At a point $B(x, F(x) + C)$?

A6: The slope of both functions is F'(x), since (F(x) + C)' = F'(x). This is also shown by the slopes of the tangent lines on the graph.

7.   Plot $y = F'(x)$.

a. Click on the expression for the slope of the tangent line and choose **Edit → Copy As →** String .

b. Choose **Draw →** Function. Select Type → Cartesian. In the Y = prompt use Ctrl-V to paste the copied expression.

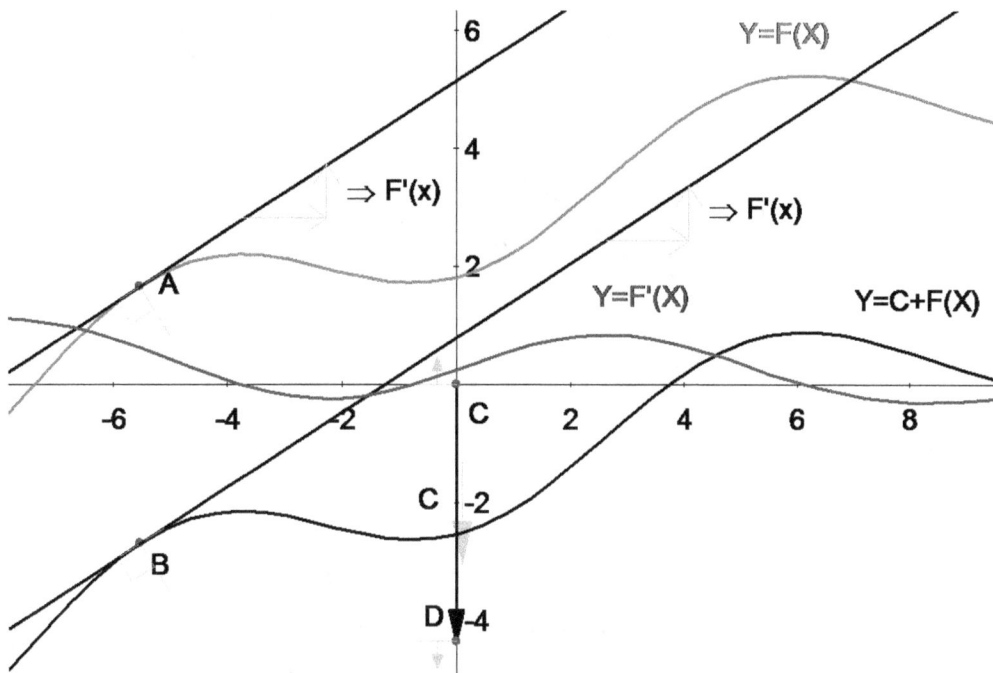

8. Drag point A along the curve of $y = F(x)$ and compare the values of $f(x) = F'(x)$ with the slope of the tangent line.

Q7. Can you justify that $G(x) = F(x) + C$ is a family of antiderivatives of $f(x) = F'(x)$?

A7: *For all values of C the graph of G(x) is "parallel" to the graph of F(x) or they have the same slope at the same point that is defined by the value of f(x).*

Note: we use the expression "parallel" graphs which, in this case, means the graph derived from the graph of the function translated by a vertical vector.

9. Plot the family of curves $F(x) + C$ using Gx's **Trace** tool.

a. Select the graph of $F(x) + C$ and choose **Construct →** Trace.

b. Select C for the parametric value and choose Start = -4, End = 4, and choose a count of your preference.

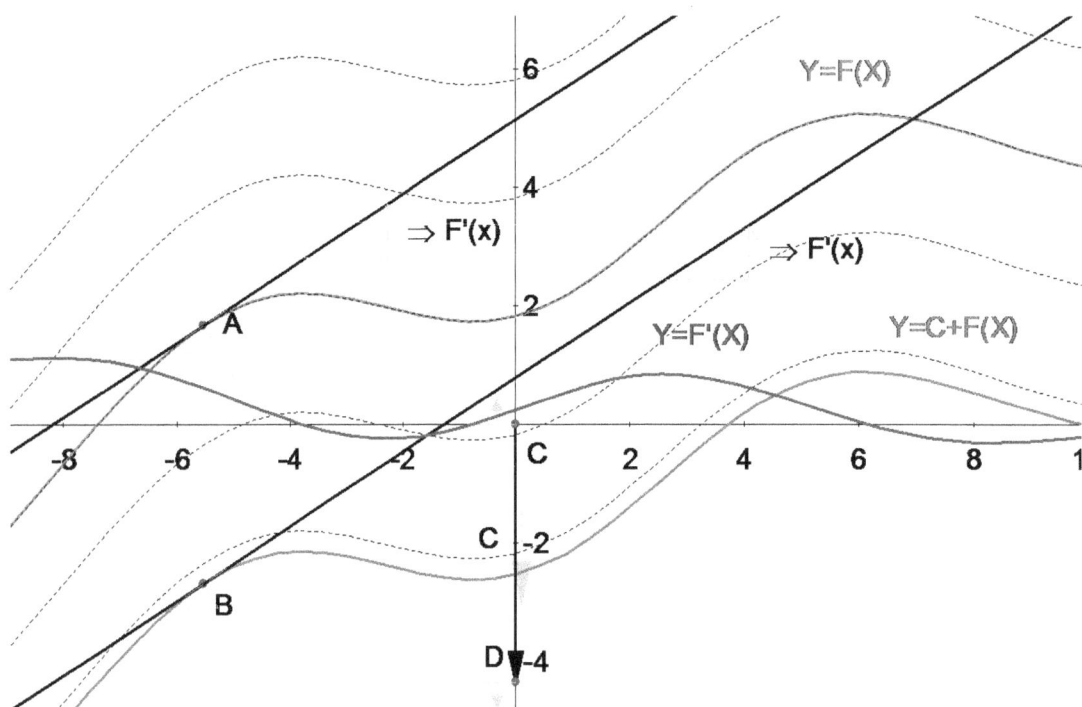

Q8. Is it possible to find a function $H(x) \neq F(x) + C$, so that $f(x) = H'(x)$?

A8. *This is impossible, since only functions whose graphs are "parallel" to the graph of F(x) have the same slope as F(x), which means $H'(x) \neq F'(x) \neq f(x)$.*

Representation of the Antiderivative

Exploration 4.1: Suppose you were asked to find a function F whose derivative is $f(x) = 2x$. From your knowledge of derivatives, you would probably say that $F(x) = x^2$, since $\frac{d(x^2)}{dx} = 2x$. Is this the only function with the given derivative? How many functions can you find with this derivative? What is the relative position of their graphs? You will explore these questions starting with a specific example and then generalizing for the function $y = f(x)$.

THE ANTIDERIVATIVE OF $f(x) = 3x^2$

1. Draw a function $f(x) = 3x^2$.

Q1. If $f(x) = \frac{dF}{dx}$, then $F(x)$ is called antiderivative of $f(x)$. What is $F(x)$?

2. Draw the antiderivative $F(x)$ on the same graph as $f(x)$.

3. Verify your formula graphically, using a tangent line.

Q2. Can you think of other functions that have the same derivatives?

Q3. How can you represent the entire family of these functions by a formula?

4. Plot antiderivative in its general form. Verify that the antiderivative in the general form has the same derivative as $F(x)$.

Note: you may want to delete or hide the graph of f(x) to unclutter your diagram.

Q4. How can you produce geometrically any antiderivative from a given antiderivative?

5. Verify your idea with the software.

ANTIDERIVAIVE OF $f(x) = \frac{dF}{dx}$

6. Replace $y = x^3$ with $y = F(x)$, and $y = x^3 + C$ with $y = F(x) + C$

Q5. Compare the graph of the general antiderivative $F(x) + C$ with the translated image of $F(x)$ by a vector $\overrightarrow{CD} = (0, C)$. Does your answer to Q4 hold for a generic function $F(x)$?

Q6. What is the slope of the tangent line at point A(x, $F(x)$)? At a point B(x, $F(x) + C$)?

7. Plot $y = F'(x)$.

8. Drag point A along the curve of $y = F(x)$ and compare the values of $f(x) = F'(x)$ with the slope of the tangent line.

Q7. Can you justify that $G(x) = F(x) + C$ is a family of an antiderivatives of $f(x) = F'(x)$?

9. Plot the family of curves $F(x) + C$ using Gx's **Trace** tool.

Q8. Is it possible to find a function $H(x) \neq F(x) + C$, so that $f(x) = H'(x)$?

4.2 The Fundamental Theorem of Calculus

Exploration 4.2: Investigate the relationship between differentiation and integration for various functions.

SUMMARY

Mathematics Objectives:

- Find an area of a plane region bounded by the graph of a positive-valued function, the x-axis, and lines $x = a$ and $x = b$.

- Discover the relationship between the definite integral of a function and its antiderivative known as the Fundamental Theorem of Calculus.

- Apply the Fundamental Theorem of Calculus to evaluate definite integrals of specific functions.

Vocabulary:

- Curvilinear trapezoid

- Antiderivative

- Integration

- Definite and indefinite integrals

Pre-requisites:

- Antiderivatives of basic functions

Problem Notes:

- Students first explore an area under a graph of a generic positive-valued function on a given interval $[a, b]$ to establish the fact that the area under the curve can be expressed as a definite integral of the function.

- Students then use the software to determine the area under the graph of several specific functions on the same interval, $[a, b]$. Using their knowledge of antiderivatives and results provided by the software, they discover that the definite integral of a function can be found as the difference between the value of the antiderivative taken at the upper limit and the value of the antiderivative taken at the lower limit. Thus, they discover the Fundamental Theorem of Calculus.

- Students test the theorem on several specific functions – using their knowledge of antiderivatives and the software for finding areas.

- Although in this exploration we tie this theorem to an area under the curve, the Fundamental Theorem shows how to use antiderivatives to evaluate definite integrals without resorting to Reimann sums and limits.

Technology skills:

- Draw: line segment, arc, polygon, function.

- Construct: tangent line

- Constrain: point proportional along an axis

- Calculate: area

Extension:

Calculate the definite integral of the function $f(x) = \sin x$ with lower limit -x and upper limit x. Explain your result geometrically.

STEP-BY-STEP INSTRUCTIONS

AREA UNDER THE CURVE

1. Draw a function $y = f(x)$, so that $f(x) > 0$ on some interval.

   a. Use **Toggle grid and axes** to show the axes without grid.

   b. Choose **Draw** → Function. Select Type → Cartesian. In the Y = prompt type f(x).

Q1. What is the area under the curve on the interval [a, b]?

A1: $A = \int_a^b f(x)dx$ - *the area is defined as a definite integral.*

2. Confirm your answer using the software. Construct a curvilinear trapezoid bounded by the curve, x-axis, $x = a$, and $x = b$, and find its area.

   a. Plot points A and B on the x-axis and points C and D on the curve.

b. Select point A and *x*-axis, and choose **Constrain** → Point proportional. Type *a* in the edit box. Select point C and the curve, and choose **Constrain** → Point proportional. Type *a* in the edit box.

c. Select point B and *x*-axis, and choose **Constrain** → Point proportional. Type *b* in the edit box. Select point D and the curve, and choose **Constrain** → Point proportional. Type *b* in the edit box.

d. Choose **Draw** → Segment and plot segments AB, AC and BD.

e. Choose **Draw** → Arc, and plot an arc CD.

f. Select points A, B, C, and D and choose **Draw** → Polygon. Select the polygon interior and choose **Calculate** → Area.

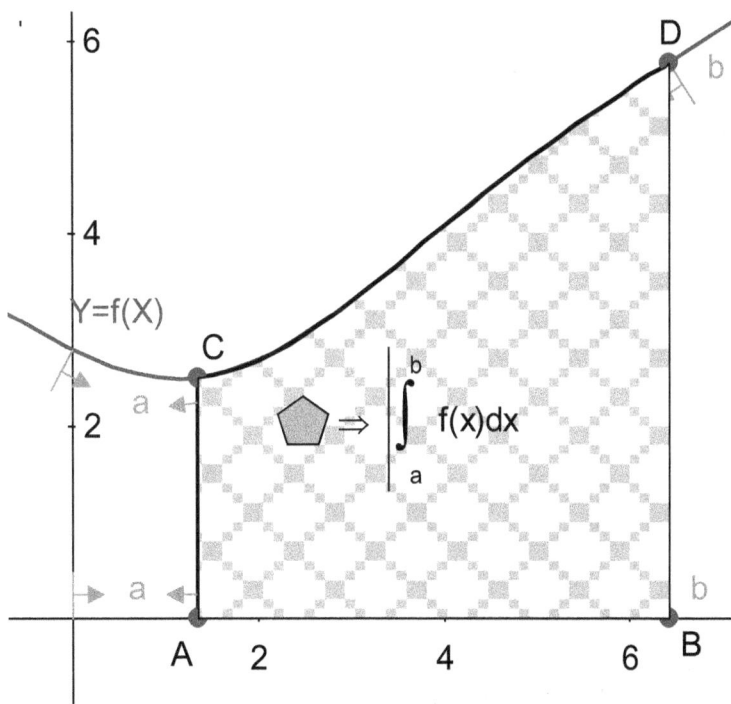

3. Find an area bounded by the graph of $f(x) = x^2$, the *x*-axis, $x = a$, and $x = b$.

   • Double-click on the expression for the function and type x^2.

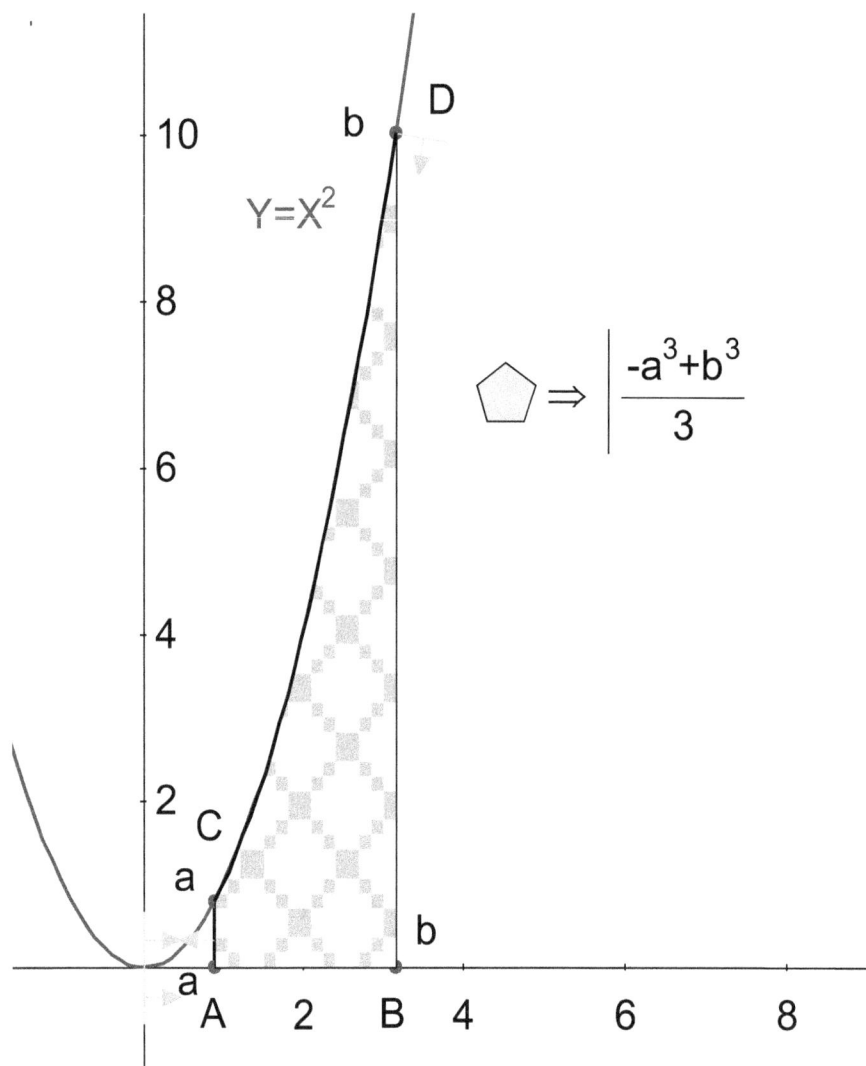

$$Y=X^2$$

$$\bigcirc \Rightarrow \left| \frac{-a^3+b^3}{3} \right|$$

Q2. What is an antiderivative of $f(x) = x^2$?

A2: An antiderivative of $f(x)=x^2$ is $F(x) = \dfrac{x^3}{3}$.

Note: encourage students to consider different particular antiderivatives and compare results of different students.

Q3. Can you see a relationship between the area of the region and the antiderivative of $f(x)$?

A3: *The area of the region is calculated as* $\int_a^b x^2 dx = \dfrac{b^3 - a^3}{3}$ *. Inserting the functions f(x) and*

F(x) into this equation we get: $\int_a^b f(x)dx = F(b) - F(a)$ *.*

4. Find an area bounded by the graph of $f(x) = \cos x$, the x-axis, $x = a$, and $x = b$.

   a. Double-click on the expression for the function and type cos(x).

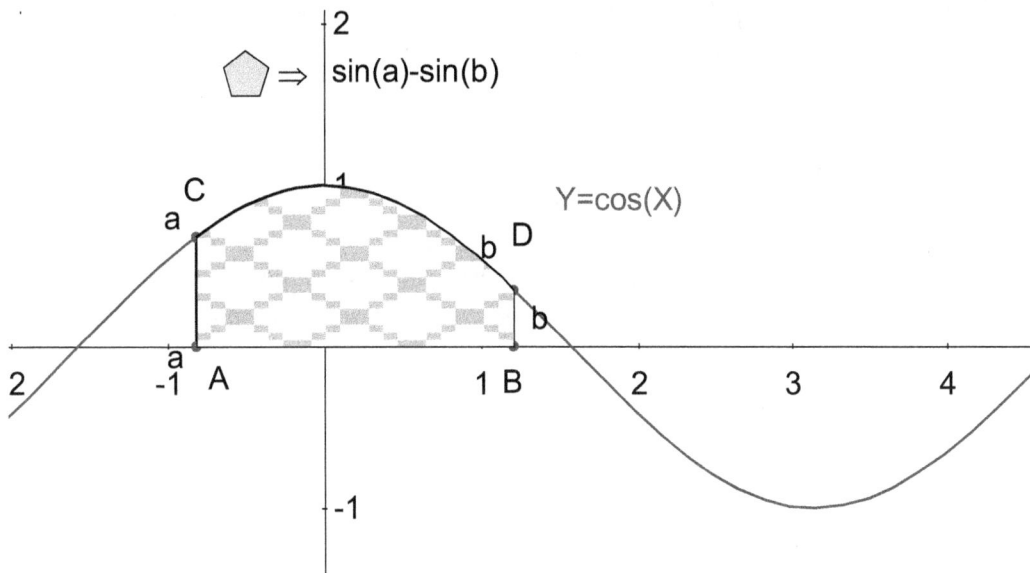

Q4. What is an antiderivative of $f(x) = \cos x$?

A4: *An antiderivative of* $f(x) = \cos x$ *is* $F(x) = \sin x$ *.*

Q5. Does your answer about the relationship between the area of the region and the antiderivative of $f(x)$ hold true in this case?

A5: *Yes, the area of the region is calculated as* $\int_a^b \cos x dx = \sin b - \sin a$ *. Inserting the*

functions f(x) and F(x) into this equation we get: $\int_a^b f(x)dx = F(b) - F(a)$ *.*

Q6. Based on your observations, complete the following statement: If a function f is continuous on the closed interval $[a, b]$ and F is an antiderivative of f on the interval $[a, b]$, then $\int_a^b f(x)dx =$ _____ .

A6: $\int_a^b f(x)dx = F(b) - F(a)$ - this is The Fundamental Theorem of Calculus.

Q7. Find the area of the region bounded by the graphs of the equations given below and verify your results using the software.

    a. $y = 1 + \sqrt{x}$, $x=1$, $x = 4$, $y = 0$

    b. $y = x + \sin x$, $x = 0$, $x = \pi$, $y = 0$.

    c. $y = e^x$, $x = 0$, $x = 2$, $y = 0$.

A7. Students can actually use a and b for the lower and upper limits of the integral and the software will provide them with a symbolic answer. If they use numeric values for the limits of integrals, they will get numeric values for the area.

    a. $\int_1^4 \left(1+\sqrt{x}\right)dx = x + \frac{2}{3}x^{\frac{3}{2}}\Big|_1^4 = 3 + \frac{14}{3} = \frac{23}{3}$

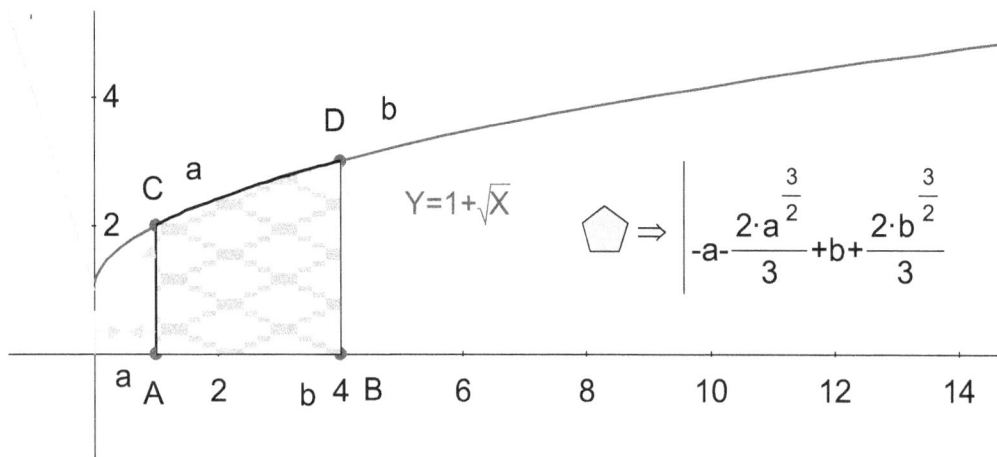

$Y = 1 + \sqrt{X}$

$\pentagon \Rightarrow \left| -a - \frac{2 \cdot a^{\frac{3}{2}}}{3} + b + \frac{2 \cdot b^{\frac{3}{2}}}{3} \right|$

b. $\displaystyle\int_0^\pi (x+\sin x)\,dx = \frac{x^2}{2} - \cos x \bigg|_0^\pi = \frac{\pi^2}{2} + 2$

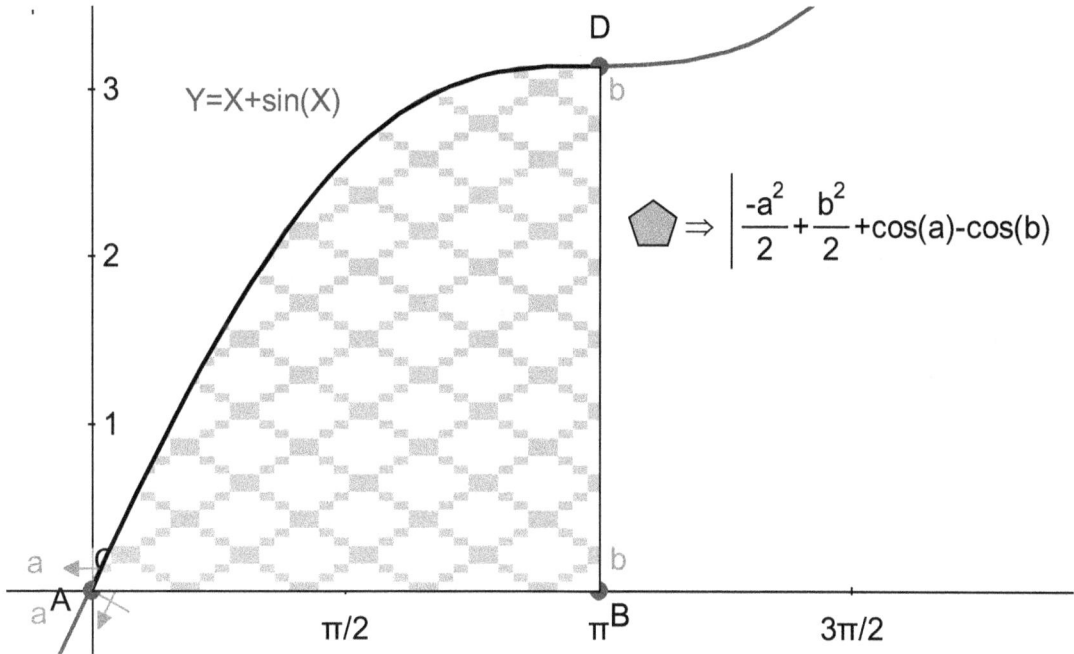

Y=X+sin(X)

$$\pentagon \Rightarrow \left| \frac{-a^2}{2} + \frac{b^2}{2} + \cos(a) - \cos(b) \right|$$

c. $\displaystyle\int_{0}^{2} e^{x} dx = e^{x}\Big|_{0}^{2} = e^{2} - 1$

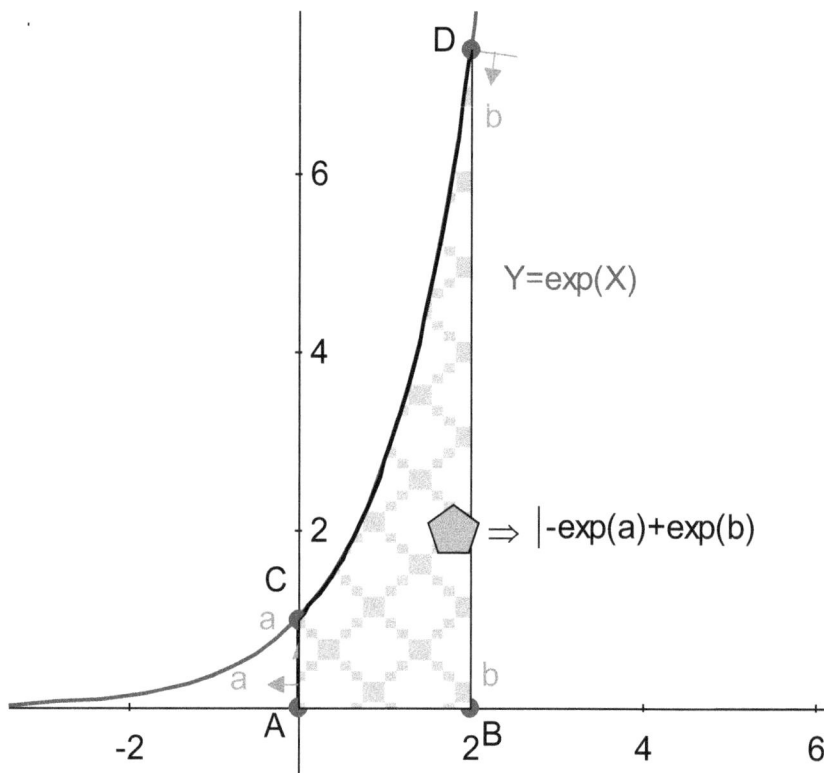

$\Rightarrow |-\exp(a)+\exp(b)|$

5. Verify your calculations using the software.

   a. Choose **Draw** → Function. Select Type → Cartesian. In the Y = prompt type the expression for a given function.

   b. For each case construct a curvilinear polygon using the instructions provided in step 2.

   c. Select the polygon interior and choose **Calculate** → Area.

   Alternatively, you can just double-click the function, enter the new one and adjust *a* and *b*. The area will be updated automatically.

The Fundamental Theorem of Calculus

Exploration 4.2: Investigate the relationship between differentiation and integration for various functions.

AREA UNDER THE CURVE

1. Draw a function $y = f(x)$, so that $f(x) > 0$ on some interval.

Q1. What is the area under the curve on the interval $[a, b]$?

2. Confirm your answer using the software. Construct a curvilinear trapezoid bounded by the curve, the x-axis, $x = a$, and $x = b$, and find its area.

3. Find an area bounded by the graph of $f(x) = x^2$, the x-axis, $x = a$, and $x = b$.

Q2. What is an antiderivative of $f(x) = x^2$?

Q3. Can you see any relationship between the area of the region and the antiderivative of $f(x)$?

4. Find an area bounded by the graph of $f(x) = \sin x$, the x-axis, $x = a$, and $x = b$.

Q4. What is an antiderivative of $f(x) = \cos x$?

Q5. Does your answer about the relationship between the area of the region and the antiderivative of $f(x)$ hold true in this case?

Q6. Based on your observations, complete the following statement: If a function f is continuous on the closed interval $[a, b]$ and F is an antiderivative of f on the interval $[a, b]$, then $\int_a^b f(x)dx = $ _____.

Q7. Find the area of the region bounded by the graphs of the equations given below and verify your results using the software.

    a. $y = 1 + \sqrt{x}$, $x=1$, $x = 4$, $y = 0$

    b. $y = x + \sin x$, $x = 0$, $x = \pi$, $y = 0$.

    c. $y = e^x$, $x = 0$, $x = 2$, $y = 0$.

4.3 The Second Fundamental Theorem of Calculus

<u>Exploration 4.3:</u> Consider the definite integral of a function $y = f(x)$ with a variable upper limit x. What is the rate of change of the integral and how does that relate to the integrand f?

SUMMARY

<u>Mathematics Objective:</u>

- Discover the 2nd Fundamental Theorem of Calculus: $\dfrac{d}{dx}\left[\displaystyle\int_a^x f(x)dx\right] = f(x)$.

- Establish the fact that differentiation and integration are inverse operations.

<u>Vocabulary:</u>
- Rate of change
- Antiderivative
- Integral
- Area as a definite integral

<u>Pre-requisites:</u>
- Derivative as a slope of the tangent line.
- Basic rules of differentiation
- Antiderivative
- Definite integral
- The Fundamental Theorem of Calculus

<u>Problem Notes:</u>

- Students explore the relationship between the slope of the tangent line to the graph of the antiderivative of $f(x)$ written as a definite integral with a variable upper limit and the function $f(x)$. They use geometric concepts of area and tangent line to establish the relationship between them. This relationship is known as the second fundamental theorem of calculus. While this theorem is often confusing to students, the geometrical representation can help them understand the meaning of this theorem: when you apply two operations that are the inverse of each other to a given function, you will get the original function.

- Students first explore the definite integral of a specific function with a variable upper limit, also known as the area function. They plot the graph of this function and analyze the rate of change of this function using the slope of the tangent line.

- Students then generalize their findings for a generic function.

Technology skills:

- Draw: function, polygon, arc, line segment

- Constrain: point proportional along the curve

- Construct: tangent line

- Calculate: area, slope

STEPS-BY-STEP INSTRUCTIONS

THE DEFINITE INTEGRAL AS A FUNCTION

1. Draw a function $y = x^2$.

   a. Use **Toggle grid and axes** to show the axes without grid.

   b. Choose **Draw** → Function. Select Type → Cartesian. In the Y = prompt type x^2.

2. Plot the region bounded by the graphs of $y = x^2$, $y = 0$, $x = a$, and a variable upper boundary x.

   a. Plot points A and B on the x-axis and points C and D on the curve.

   b. Select point A and x-axis, and choose **Constrain** → Point proportional. Type a in the edit box. Select point C and the curve, and choose **Constrain** → Point proportional. Type a in the edit box.

   c. Select point B and x-axis, and choose **Constrain** → Point proportional. Type x in the edit box. Select point D and y-axis, and choose **Constrain** → Point proportional. Type x in the edit box.

   d. Choose **Draw** → Segment and plot segments AB, AC and BD. Choose **Draw** → Arc, and plot an arc CD.

e. Select points A, B, C, and D and choose **Draw** → Polygon.

Q1. What is the area of this plane region?

A1: $A(x) = \int\limits_{a}^{x} t^2 dt = \dfrac{x^3}{3} - \dfrac{a^3}{3}$.*The area is a function of x, a = const.*

2. Verify your answer with the help of software.

- Select the polygon interior and choose **Calculate** → Area.

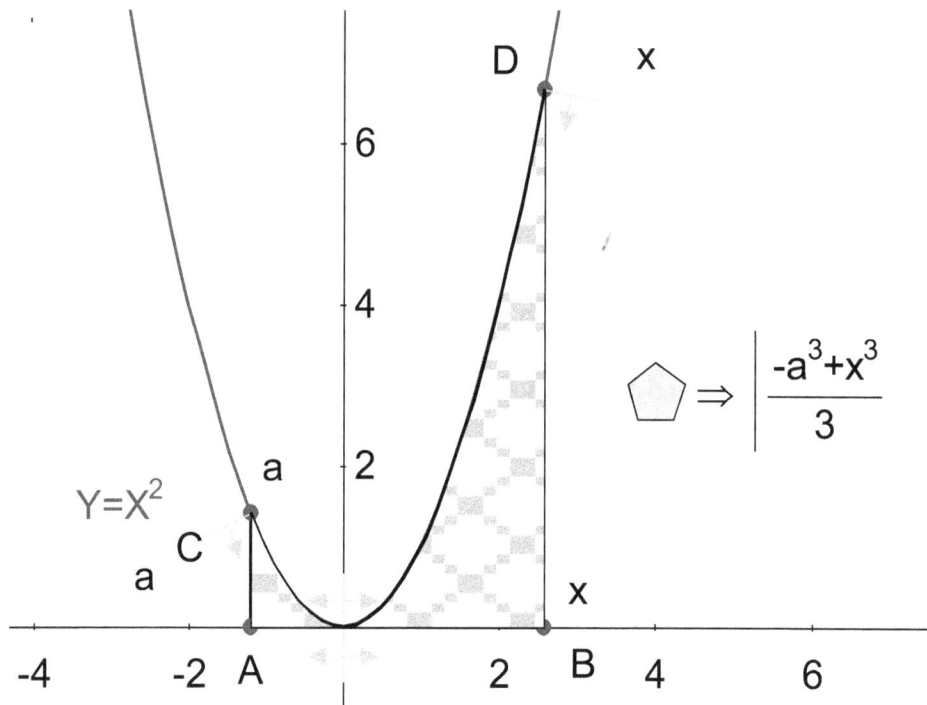

THE RATE OF CHANGE OF THE AREA FUNCTION

Q2. The area of the region is a function of *x*. How does the area change as *x* changes? What is the derivative of this function?

A2: As x increases, the area increases. $\dfrac{d}{dx}\left(\dfrac{x^3 - a^3}{3}\right) = x^2$.

3. To check the answer to question 2, plot the graph of the area of the region bounded by the graphs of $y = x^2$, $y = 0$, $x = a$, and a variable upper boundary x. Find the slope of this graph at a point x.

    a. Click on the expression for the area and choose **Edit** → Copy As → String.

    b. Choose **Draw** → Function. Choose Cartesian for Type and paste the area function in the Y= prompt using Ctrl-V. Delete "abs" in the expression of the function. (Or, before copying the string, set the **Output Properties** →Use Assumptions →Yes from the right-click context menu. This gets rid of the absolute value symbols.)

    c. Select the graph of the area function and choose **Construct** → Tangent to curve.

    d. Select the point of tangency (point E) and the curve and choose **Constrain** → Point proportional. In the open edit box type x.

    e. Select the tangent line and choose **Calculate** → Slope.

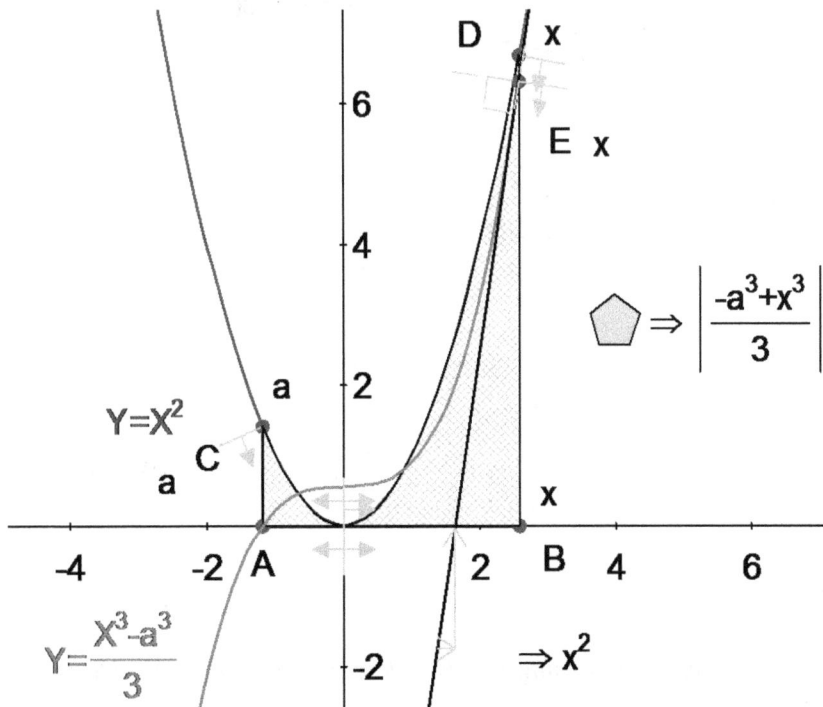

Q3. How does the slope of the tangent line to the area function compare to the original function that bounds the region?

A3: They are equal.

Q4. Will your answer hold true for other functions? Explain your answer.

A4: Answers will vary.

4.   Verify your decision by choosing another function and repeating the work above.

   a.   Double-click on the expression Y = x^2 and type a different function.

   b.   Click on the expression for the area and choose **Edit** → Copy As → String.

   c.   Choose **Draw** → Function. Choose Cartesian for Type and paste the area function in the Y= prompt using Ctrl-V. Delete "abs" in the expression of the function.

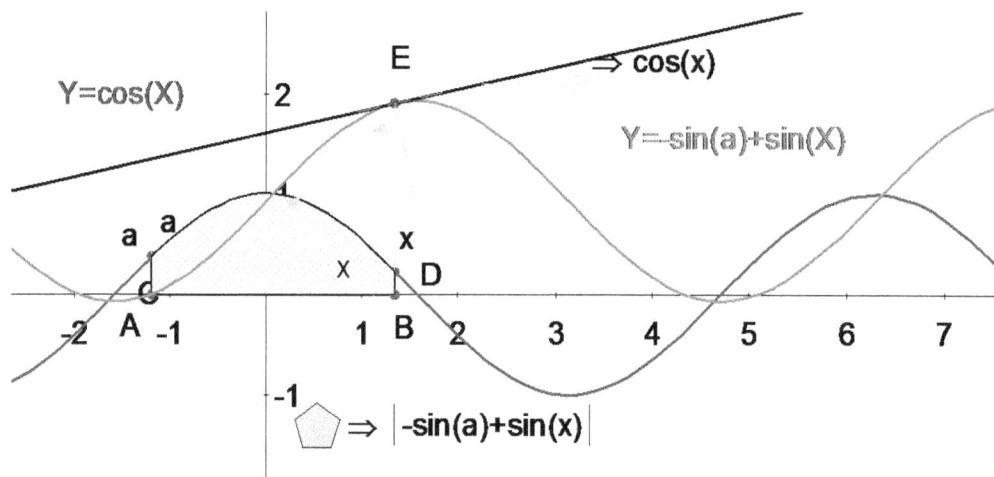

Q5. Write a general expression for the area of the plane region between the graph of the function $y = f(x)$ and x-axis on the interval $[a, x]$.

A5: $A(x) = \int_{a}^{x} f(t)\,dt$

Note: to avoid the confusion of using x in two different ways, the variable of integration is switched to t.

Q6. Write a general expression for the derivative of the area function.

A6: $\dfrac{dA}{dx} = f(x)$.

Q7. Use your answers to questions 5 and 6 and complete the following statement: If *f* is continuous on an open interval containing *a*, then, for every *x* in the interval,

$$\frac{d}{dx}\left[\int_a^x f(x)dx\right] = \underline{\hspace{3cm}}.$$

A7: $\dfrac{d}{dx}\left[\displaystyle\int_a^x f(x)dx\right] = f(x)$ - this is The 2nd Fundamental Theorem of Calculus.

The Second Fundamental Theorem of Calculus

Exploration 4.3: Consider the definite integral of a function $y = f(x)$ with a variable upper limit x. What is the rate of change of the integral and how does that relate to the integrand f?

THE DEFINITE INTEGRAL AS A FUNCTION

1. Draw a function $y = x^2$.

2. Plot the region bounded by the graphs of $y = x^2$, $y = 0$, $x = a$, and a variable upper boundary x.

Q1. What is the area of this plane region?

2. Verify your answer with the help of the software.

THE RATE OF CHANGE OF THE AREA FUNCTION

Q2. The area of the region is a function of x. How does the area change as x changes? What is the derivative of this function?

3. To check the answer to question 2, plot the graph of the area of the region bounded by the graphs of $y = x^2$, $y = 0$, $x = a$, and a variable upper boundary x. Find the slope of this graph at a point x.

Q3. How does the slope of the tangent line to the area function compare to the original function that bounds the region?

Q4. Will your answer hold true for other functions? Explain your answer.

4. Verify your decision by choosing another function and repeating the work above.

Q5. Write a general expression for an area of the plane region between the graph of the function $y = f(x)$ and x-axis on the interval $[a, x]$.

Q6. Write general expression for the derivative of the area function.

Q7. Use your answers to questions 5 and 6 and complete the following statement: If f is continuous on an open interval containing a, then, for every x in the interval,

$$\frac{d}{dx}\left[\int_a^x f(x)dx \right] =$$

4.4 Integral of an Inverse Function

Exploration 4.4: In this problem you will explore a method to calculate integrals of functions by using the integral of their inverse function. This method can be used to integrate functions that otherwise require very special techniques of integration.

First explore a case with a simple function $y = \sqrt{x}$ so you can verify the method by direct integration. Then apply this method to a function $y = \arcsin x$. This function requires a special technique in order to integrate it directly.

SUMMARY

Mathematics Objectives:

- Compare integrals of functions and their inverses.

- Explore the method of integration of functions using the integrals of their inverses.

Vocabulary:

- Curvilinear polygon

- Rigid transformation

Pre-requisites:

- Inverse functions

- Reflection

- Integrals of basic functions

Problem Notes:

- This exploration provides students with an alternate method of finding definite integrals based on the idea that the area bounded by the graph of a function and the x-axis is equal to the area bounded by the graph of the inverse of the function and the y-axis when limits of integration correspond to each other. The equality of areas follows from the symmetry of graphs of a function and its inverse relative to the identity line $y = x$.

- Students first explore the area of plane region bounded by the graph of a function and the x-axis and compare it to the area of a region bounded by the graph of the inverse of the function and the *y*-axis. From symmetry considerations, they determine the algebraic relationship between the definite integrals, and thus develop an alternate method of integration of functions.

- Students then apply the method of integration they found to a more complex function that would otherwise require integration by parts. They confirm their calculations using the software.

Technology skills:

- Draw: function, point, line, segment, arc, polygon

- Constrain: point proportional along the curve, slope

- Calculate: area, implicit equation

- Construct: reflection

Extensions:

1. Find $\int_{1}^{a} \ln x\, dx$ using the inverse function. (Note: this problem illustrates the method of integration by parts).

2. Given $y = f(x)$ that has an inverse $g(x) = f^{-1}(x)$ on an interval $[a,b]$. Setup a definite integral $\int_{f(a)}^{f(b)} g(x)\, dx$ in terms of $\int_{a}^{b} f(x)\, dx$.

STEP-BY-STEP INSTRUCTIONS

INTEGRAL OF $f(x)=\sqrt{x}$ ON [4, 9]

1. Draw a function $y = \sqrt{x}$

    a. Use **Toggle grid and axes** to show the axes without a grid.

    b. Choose **Draw** → Function. Select Type → Cartesian. In the Y = prompt type sqrt(x).

2. Construct a curvilinear polygon bounded by the graphs of the equations $y = \sqrt{x}$, $y = 0$, $x = 4$, and $x = 9$.

    a. Choose **Draw** → Point and plot points A and B on the *x*-axis, and points C and D on the curve.

    b. Select point A and the *x*-axis, and choose **Constrain** → Point proportional. Type 4 in the open edit box.

    c. Select point C and the curve, and choose **Constrain** → Point proportional. Type 4 in the open edit box.

    d. Select point B and the *x*-axis, and choose **Constrain** → Proportional. Type 9 in the open edit box.

    e. Select point D and the curve, and choose **Constrain** → Proportional. Type 9 in the open edit box.

    f. **Draw** →Line Segment(s) AC, AB, and BD.

    g. Choose **Draw** → Arc and plot the arc CD.

    h. Select the arc CD and the segments AB, AC, and BD. Choose **Draw** → Polygon.

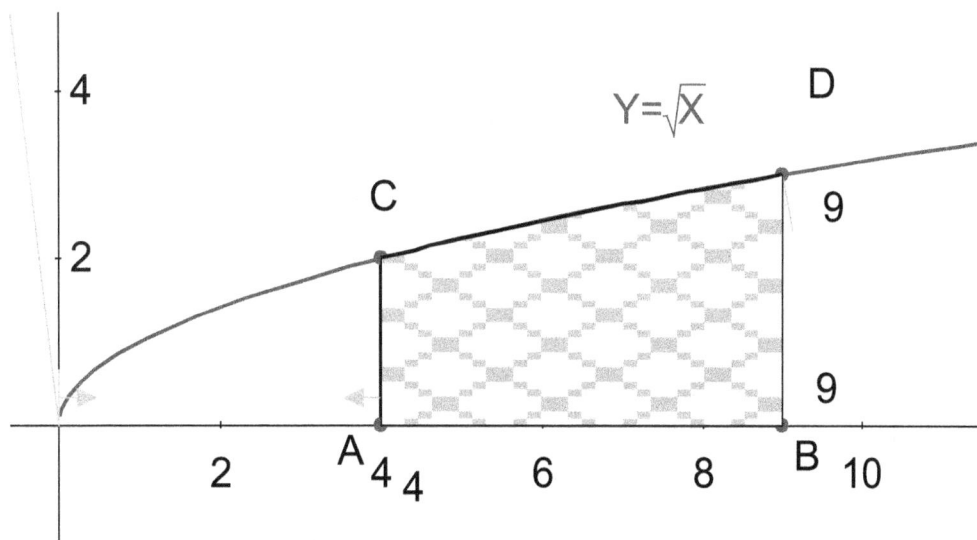

Q1. What is the area of the curvilinear polygon ABDC? Setup an integral and find the area.

A1: $\int_{4}^{9} \sqrt{x}\,dx = \frac{2}{3}x^{\frac{3}{2}}\Big|_{4}^{9} = \frac{38}{3}$

3. Verify your formula using the software.

    a. Select the polygon interior and choose **Calculate** → Area.

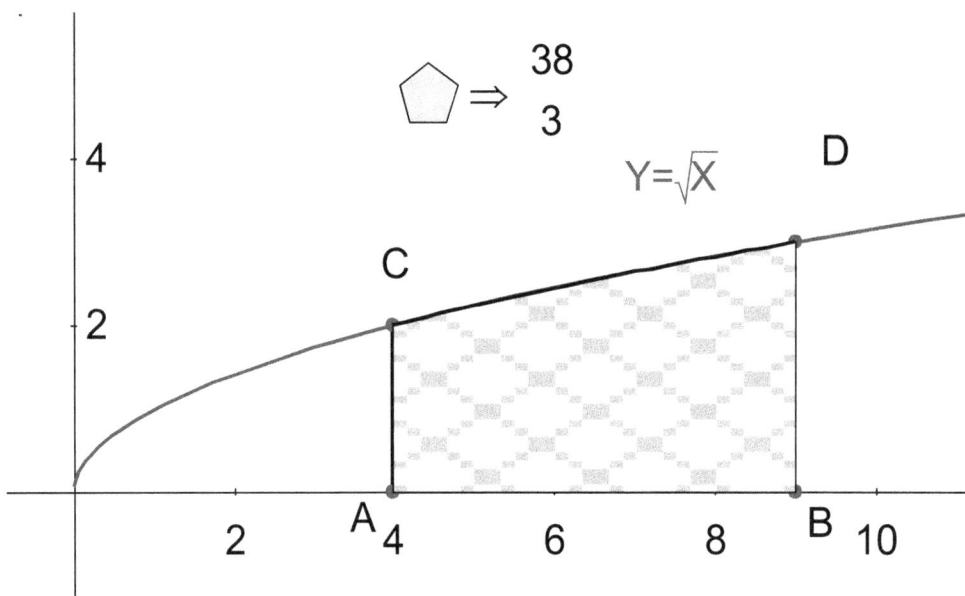

Q2. Will the area of the curvilinear polygon be the same if you reflect the region about the line $y = x$?

A2: Since reflection is a rigid transformation, the area will be exactly the same.

Q3. What function is produced by the reflection of the graph $y = \sqrt{x}$ about the line $y = x$?

A3: This will be the inverse function to $y = \sqrt{x}$, so $x = \sqrt{y}$ or $y = x^2$ for $y > 0$.

4. Use the software to reflect the function $y = \sqrt{x}$ over the line y=x. To verify your answer in Q3, find an equation for the reflected graph.

    a. Plot a point at the origin. Choose **Draw** → Line going through the origin point. (Hint: when plotting the point, make sure you have both bowtie makes over the cursor when you click the point. It may look like it's at the origin, but if you can drag the point, then it's not constrained. If you have difficulty, just use the **Constrain** → Coordinate tool.)

b. Select the line and choose **Constrain** → Slope. Type 1 in the open edit box.

c. Select the graph of $y = \sqrt{x}$ and choose **Construct** → Reflection. Then click on the line of reflection.

d. Select the reflected curve and choose **Calculate** → Implicit Equation.

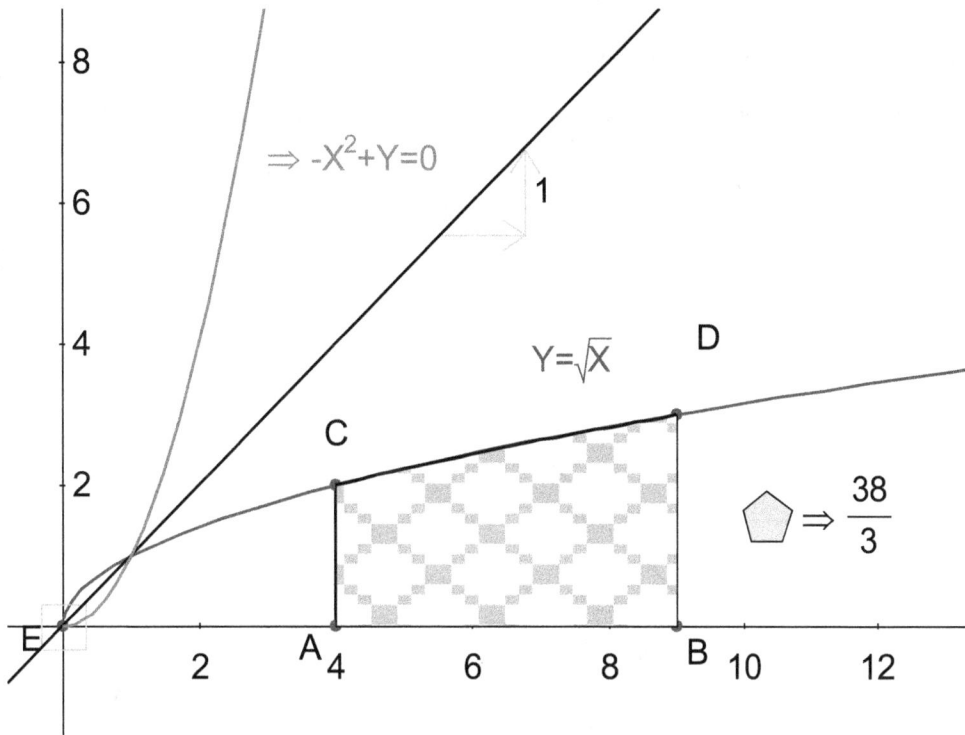

5. Construct a curvilinear polygon that is the reflected image of the curvilinear polygon ABDC.

a. Choose **Draw** → Point and plot points F and G on the *y*-axis, and points H and I on the reflected curve.

b. Select point F and the *y*-axis, and choose **Constrain** → Proportional. Type 4 in the open edit box.

c. Select point H and the curve, and choose **Constrain** $\rightarrow$ Proportional. Type 4 in the open edit box.

d. Select point G and the *y*-axis, and choose **Constrain** $\rightarrow$ Proportional. Type *9* in the open edit box.

e. Select point I and the curve, and choose **Constrain** $\rightarrow$ Proportional. Type *9* in the open edit box.

f. **Draw** $\rightarrow$Line Segment(s) FH, FG, and GI.

g. Choose **Draw** $\rightarrow$ Arc and plot an arc HI.

h. Select arc HI and segments FG, FI, and GH, and choose **Construct** $\rightarrow$ Polygon.

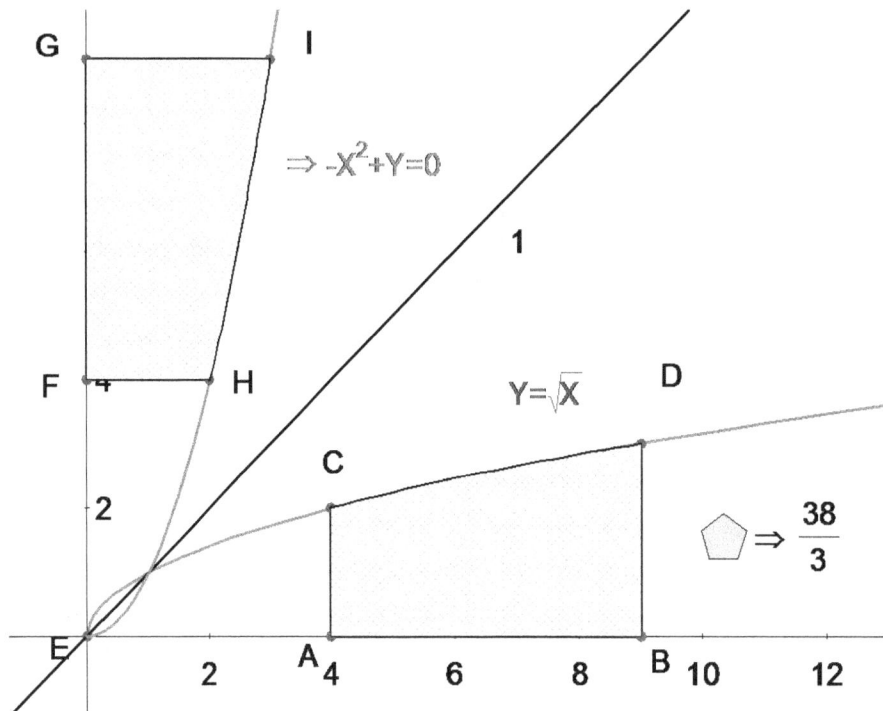

Q4. How can you find the area of the curvilinear polygon FGIH?

A4: *Based on the graph, the area of the region is equal to the area of a rectangle GFHP between the lines y = 9 and y = 4 on [0, 2] and area of a region between the line y = 9 and y = x^2 on [2, 3]:*

$$A = 2\cdot(9-4) + \int_2^3 (9-x^2)\,dx$$

$$= 10 + \left(9x - \frac{x^3}{3}\right)\Bigg|_2^3 = 10 + 9(3-2) - \frac{3^3}{3} + \frac{2^3}{3} = 19 - \frac{19}{3} = \frac{38}{3}$$

which is the same area as ACDB.(See the diagram below.)

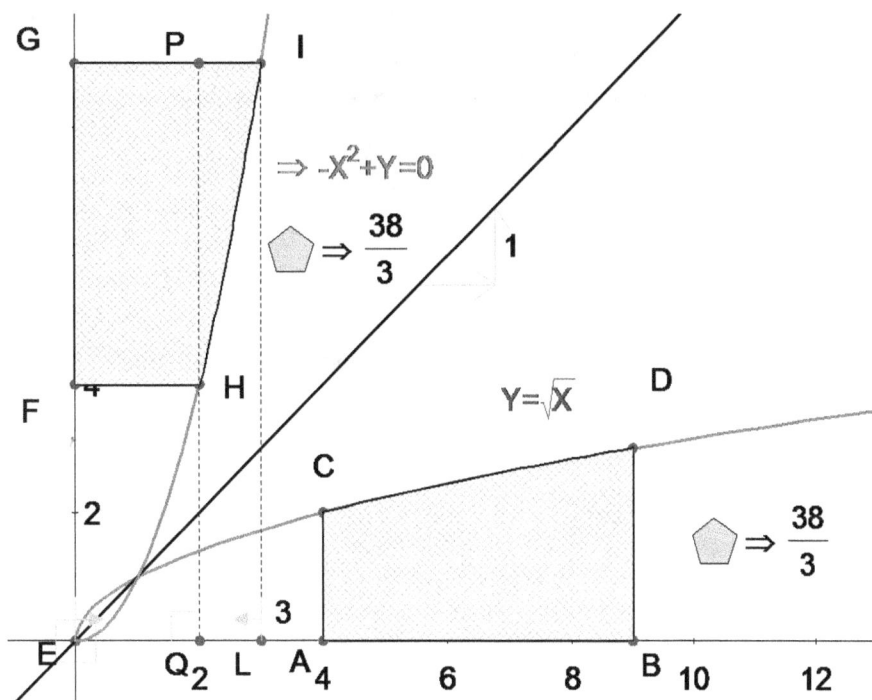

THE INTEGRAL OF $f(x) = \arcsin x$ ON $[0, 1]$

Q5. Apply the same idea to the integral of $y = \arcsin x$, and setup a formula for the area of the region bounded by the graphs of the equations: $y = \arcsin x$, $y = 0$, $x = 0$, and $x = 1$

A5: $\displaystyle \int_0^1 \arcsin x\,dx = \frac{\pi}{2} - \int_0^{\frac{\pi}{2}} \sin x\,dx = \frac{\pi}{2} + \cos x\Big|_0^{\frac{\pi}{2}} = \frac{\pi}{2} - 1$

6. Verify your result with the help of the software

   a. Open a new document. Use **Toggle grid and axes** to show the axes without a grid.

   b. Choose **Draw** → Function. Select Type → Cartesian. In the Y = prompt type arcsin(x).

7. Construct a curvilinear polygon bounded by the graphs of equations $y = \arcsin x$, $y = 0$, $x = 0$, and $x = 1$. In Gx, curves in curvilinear polygons must be connected with line segments, even if the segments have a length of 0. For that reason, it's important to construct the polygon first, then constrain the points, like this:

   a. Choose **Draw** → Point and plot points A and B on the *x*-axis, and points C and D on the curve. Note: don't put A or C at the origin.

   b. **Draw** →Line Segment(s) AC, AB, and BD.

   c. Choose **Draw** → Arc CD.

   d. Select the arc CD and the segments AB, AC, and BD. Choose **Draw** → Polygon.

   e. Select point A and the *x*-axis, and choose **Constrain** → Point proportional. Type 0 in the open edit box.

   f. Select point C and the curve, and choose **Constrain** → Point proportional. Type 0 in the open edit box.

   g. Select point B and the *x*-axis, and choose **Constrain** → Point proportional. Type 1 in the open edit box.

   h. Select point D and the curve, and choose **Constrain** → Point proportional. Type 1 in the open edit box.

   i. Select the polygon interior and choose **Calculate** → Area.

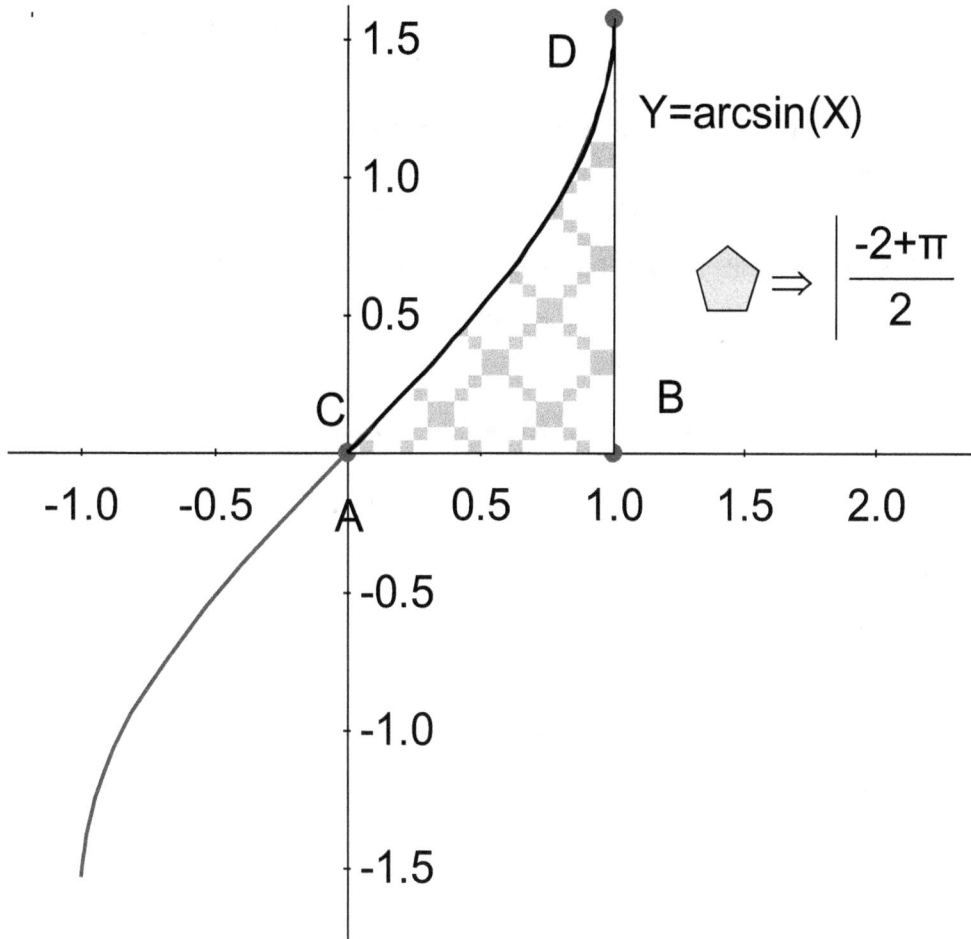

$$Y=\arcsin(X)$$

$$\pentagon \Rightarrow \left| \frac{-2+\pi}{2} \right|$$

Integral of an Inverse Function

Exploration 4.4: In this problem you will explore a method to calculate integrals of functions by using the integral of their inverse function. This method can be used to integrate functions that otherwise require very special techniques of integration.

First explore the case with a simple function $y = \sqrt{x}$ so you can verify the method by direct integration. Then apply this method to a function $y = \arcsin x$. This function requires a special technique in order to integrate it directly.

THE INTEGRAL OF $f(x) = \sqrt{x}$ ON [4, 9]

1. Draw a function $y = \sqrt{x}$

2. Construct a curvilinear polygon bounded by the graphs of the equations $y = \sqrt{x}$, $y = 0$, $x = 4$, and $x = 9$.

Q1. What is the area of the curvilinear polygon ABDC? Setup an integral and find the area.

3. Verify your formula using the software.

Q2. Will the area of the curvilinear polygon be the same if you reflect the region about the line $y = x$?

Q3. What function is produced by the reflection of the graph $y = \sqrt{x}$ about the line $y = x$?

4. Use the software to reflect the function $y = \sqrt{x}$ over the line y=x. To verify your answer in Q3, find an equation for the reflected graph.

5. Construct a curvilinear polygon that is the reflected image of the curvilinear polygon ABDC.

Q4. How can you find the area of the curvilinear polygon FGIH?

THE INTEGRAL OF $f(x) = \arcsin x$ ON [0, 1]

Q5. Apply the same idea to the integral of $y = \arcsin x$, and setup a formula for the area of the region bounded by the graphs of equations: $y = \arcsin x$, $y = 0$, $x = 0$, and $x = 1$

6. Verify your result with the help of software

7. Construct a curvilinear polygon bounded by the graphs of equations $y = \arcsin x$, $y = 0$, $x = 0$, and $x = 1$.

4.5 The Trapezoidal Method

<u>Exploration 4.5:</u> Consider a plane region bounded by the graphs of equations $y = f(x)$, $y = 0$, $x = a$, and $x = b$, where $f(x)$ is continuous on $[a, b]$. You need to evaluate the area of the region. One way to approximate the area is to use n right angled trapezoids with equal heights determined by $h = \dfrac{b-a}{n}$. Your task is to develop the Trapezoidal Method for approximating the area and determine when the approximation is an over or an under estimation.

SUMMARY

<u>Mathematics Objectives:</u>

- Develop the Trapezoidal Method for approximating the area bounded by the graphs of equations $y = f(x)$, $y = 0$, $x = a$, and $x = b$, where $f(x)$ is continuous on $[a, b]$.

- Determine when the trapezoidal approximation is an over or an under estimation using concavity of a function.

<u>Vocabulary:</u>

- Continuity

- Polygonal chain

<u>Pre-requisites:</u>

- Area of trapezoid

- Sigma notation

<u>Problem Notes:</u>

- There are a wide range of methods available for numerical integration. The most straightforward numerical integration techniques approximate a function tabulated at a sequence of regularly spaced intervals by various degree polynomials. If the endpoints are tabulated, then the 2- and 3-point formulas are called the Trapezoidal Rule and Simpson's rule, respectively.

- In this exploration students will develop the Trapezoidal Method for a generic function. They draw a polygonal chain to approximate the function on a given interval and find the sum of areas of trapezoids that approximates the area under the curve. The symbolic feature of *Gx* shows the summation process of the trapezoidal areas correlating with the students algebra work. This is a powerful reinforcement of these skills.

- Then students explore specific examples to make the connection between the concavity of the function and the type of estimation. They determine that if the curve is concave up, the chords that define the leg of the trapezoid are above the curve, hence we have an over estimate. Similarly, if the curve is concave down we have an under estimate.

Technology skills:

- Draw: function, point, curve approximation, line segment

- Constrain: point proportional along a curve, perpendicular

- Construct: polygon

- Calculate: area

Extension:

Find an approximation of the integral $\int_0^1 x^3 dx$, using the Trapezoidal Method with n trapezoids of equal heights. What is the smallest number of trapezoids n such that the result of the approximation differs from the value of the integral by less than 1%?

STEP-BY-STEP INSTRUCTIONS

TRAPEZOIDAL METHOD

1. Draw a function $y = f(x)$.

    a. Use **Toggle grid and axes** to show the axes without the grid.

    b. Choose **Draw** → Function. Choose Cartesian for Type and in the Y= prompt type f(x).

2. Approximate the graph of $y = f(x)$ from $x = a$ to $x = b$ with a polygonal chain of 4 segments. Construct a polygon containing the polygonal chain that approximates the plane region.

    a. Choose **Draw** → Point, and plot points A and B on the *x*-axis. Plot points C and D on the graph of $f(x)$.

    b. Select point C and the graph of $f(x)$, and choose **Constrain** → Point proportional. In the open edit box type *a*.

    c. Select point D and the graph of $f(x)$, and choose **Constrain** → Point proportional. In the open edit box type *b*.

    d. Choose **Draw** → Curve Approximation, and draw a polygonal chain from point C to point D. In the open edit box type 4 for the number of segments.

    e. Choose **Draw** → Line Segment, and plot segments $\overline{AC}$, $\overline{BD}$, and $\overline{AB}$.

    f. Select segment $\overline{AC}$ and the *x*-axis, and choose **Constrain** → Perpendicular.

    g. Select segment $\overline{BD}$ and the *x*-axis, and choose **Constrain** → Perpendicular.

    h. Select segments $\overline{AC}$, $\overline{BD}$, and $\overline{AB}$, and all segments of the broken line and choose **Construct** → Polygon.

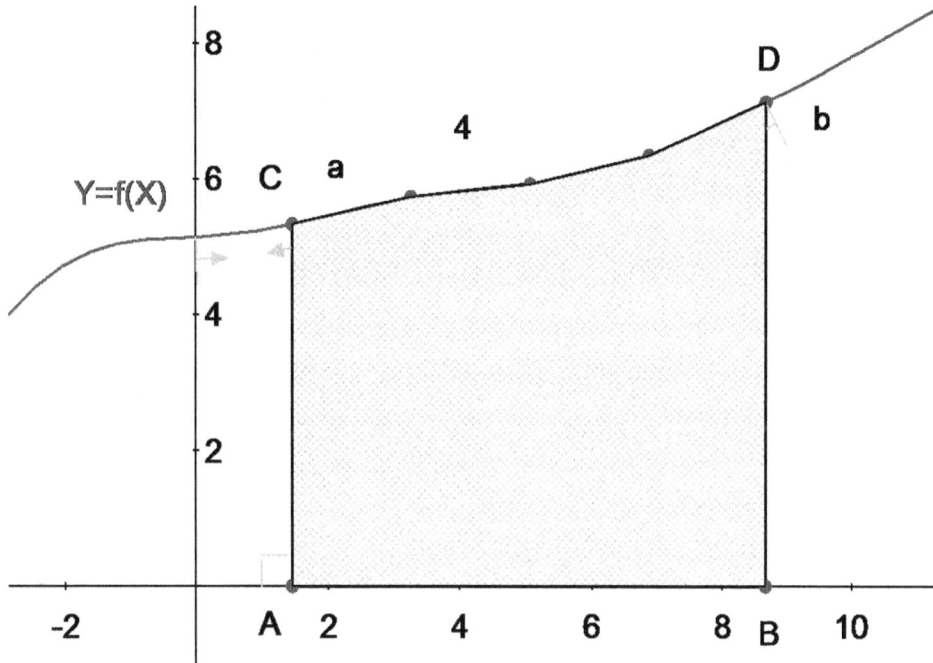

Q1. What is the area of the region bounded by the curve approximated with the 4 line segments, $y = 0$, $x = a$, and $x = b$?

A1: $A_4 = \dfrac{f(a)+f(x_1)}{2}h + \dfrac{f(x_1)+f(x_2)}{2}h + \dfrac{f(x_2)+f(x_3)}{2}h + \dfrac{f(x_3)+f(b)}{2}h$

$= \dfrac{h}{2}\big(f(a)+2f(x_1)+2f(x_2)+2f(x_3)+f(b)\big),$ where $h = \dfrac{b-a}{4}$, $x_1 = a+h$,

$x_2 = a+2h$, $x_3 = a+3h$.

3. Verify your calculations using the software.

   a. Select polygon interior and choose **Calculate** → Area.

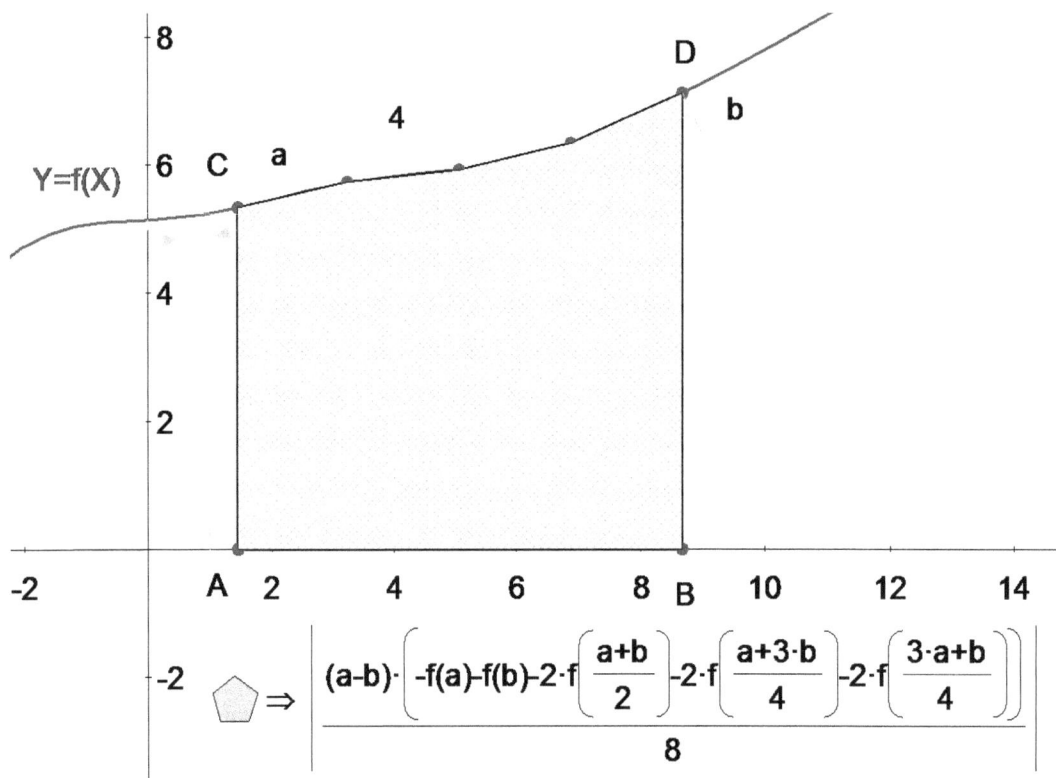

$$(a-b)\cdot\left(-f(a)-f(b)-2\cdot f\left(\frac{a+b}{2}\right)-2\cdot f\left(\frac{a+3\cdot b}{4}\right)-2\cdot f\left(\frac{3\cdot a+b}{4}\right)\right)$$
$$\over 8$$

Note: Students can easily verify that the formula they provided is equivalent to the formula provided by the software.

Q2. What is the formula for the area if you approximate the curve with a polygonal chain of *n* segments?

A2: $A_n = \dfrac{h}{2}\left(f(a)+2f(x_1)+2f(x_2)+\ldots+2f(x_{n-1})+f(b)\right)$, *where* $h = \dfrac{b-a}{n}$, $x_k = a+hk$, $k = 1,2,\ldots,n-1$

4. Verify your calculations using the software.

   a. Delete the constructed polygon.

   b. Double-click on the number of sides and type *n* in the open edit box.

   c. Select all vertices, (A through H) that form the polygon and choose **Construct →** **Polygon**.

   d. Select the polygon interior and choose **Calculate →** Area.

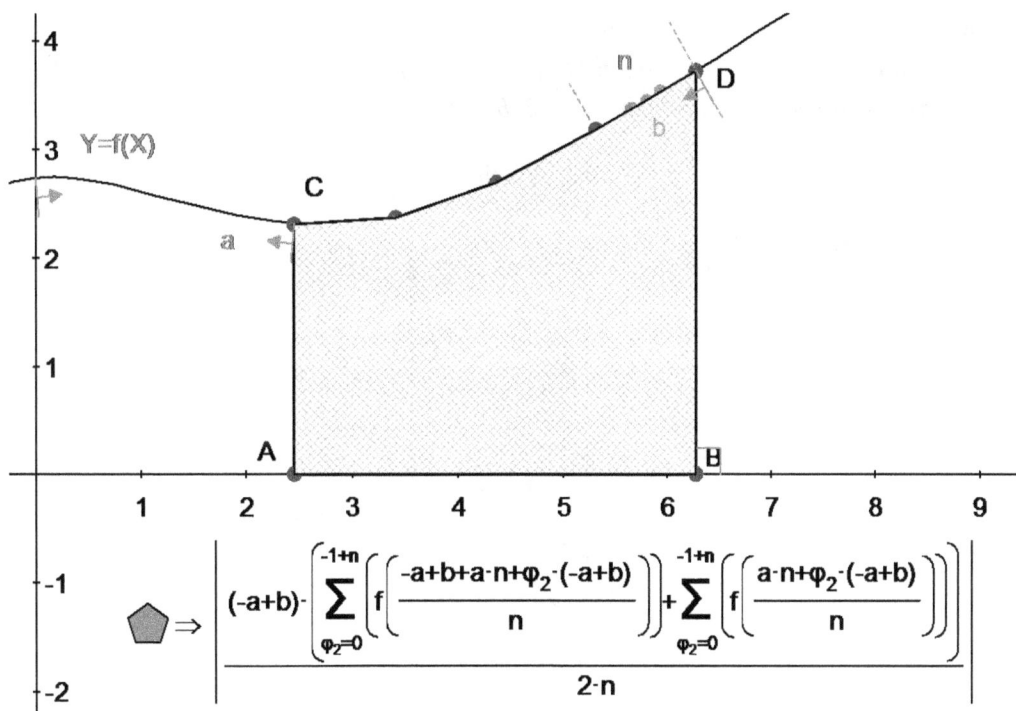

$$\left[\bigstar\right] \Rightarrow \left| \frac{(-a+b)\cdot\left[\displaystyle\sum_{\varphi_2=0}^{-1+n}\left(f\left(\frac{-a+b+a\cdot n+\varphi_2\cdot(-a+b)}{n}\right)\right)\right] + \displaystyle\sum_{\varphi_2=0}^{-1+n}\left(f\left(\frac{a\cdot n+\varphi_2\cdot(-a+b)}{n}\right)\right)}{2\cdot n} \right|$$

Note: students should verify that the formula provided by the software is equivalent to the formula they provided.

OVER AND UNDER

Q3. What do you need to know to determine if the estimated value of the area is larger or smaller than the actual value of the area?

A3: Answers will vary. Students will be able to provide a specific answer after exploring specific cases. In general, the concavity of the function determines if the estimation with trapezoids under or over estimates the area under the curve.

Q4. Let $f(x) = x^2 + 1$. What is the approximated value of $\displaystyle\int_{-2}^{2}(x^2+1)dx$ using the Trapezoidal Method with $n = 4$? Show your calculations and confirm them with the help of software.

A4: $\displaystyle\int_{-2}^{2}(x^2+1)dx \approx \frac{2-(-2)}{2\cdot4}\left(((-2)^2+1)+2((-1)^2+1)+2(0^2+1)+2(1^2+1)+(2^2+1)\right)=10$.

In order to verify this result with the software, double-click on Y = f(x) and enter the expression for the given function. Using the methods described, enter the following information given in the problem: a = -2, b = 2. n = 4.

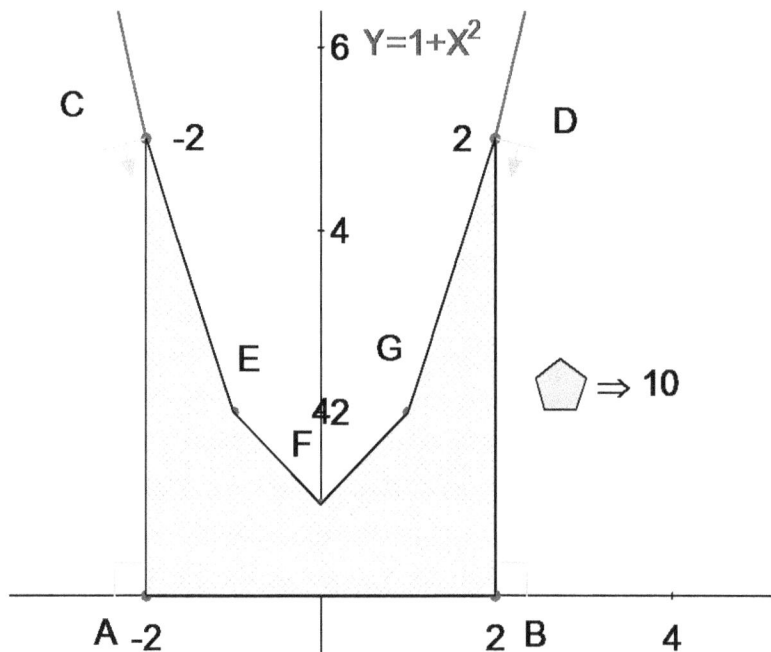

Q5. Is this an over or an under estimation? Why?

A5: *The chords are above the curve, so the approximation is an over-estimation. This can be confirmed by calculating the actual value of the area under the parabola:*

$$\int_{-2}^{2} (x^2 + 1)dx = \frac{x^3}{3} + x\Big|_{-2}^{2} = \frac{28}{3} < 10.$$

Q6. Let $f(x) = \sqrt{x}$. What is the approximated value of $\int_{1}^{9} \sqrt{x}\,dx$ using the Trapezoidal Method with $n = 4$?

6. Show your calculations and confirm them with the help of the software.

A6: $\int_{1}^{9} \sqrt{x}\,dx \approx \frac{9-1}{2 \cdot 4}\left(\sqrt{1} + 2\sqrt{3} + 2\sqrt{5} + 2\sqrt{7} + \sqrt{9}\right) \approx 17.2$

In order to verify this result with the software, double-click on Y = f(x) and enter the expression for the given function. Using the methods described, enter the following information given in the problem: $f(x) = \sqrt{x}$, $a = 1$, $b = 9$, $n = 4$.

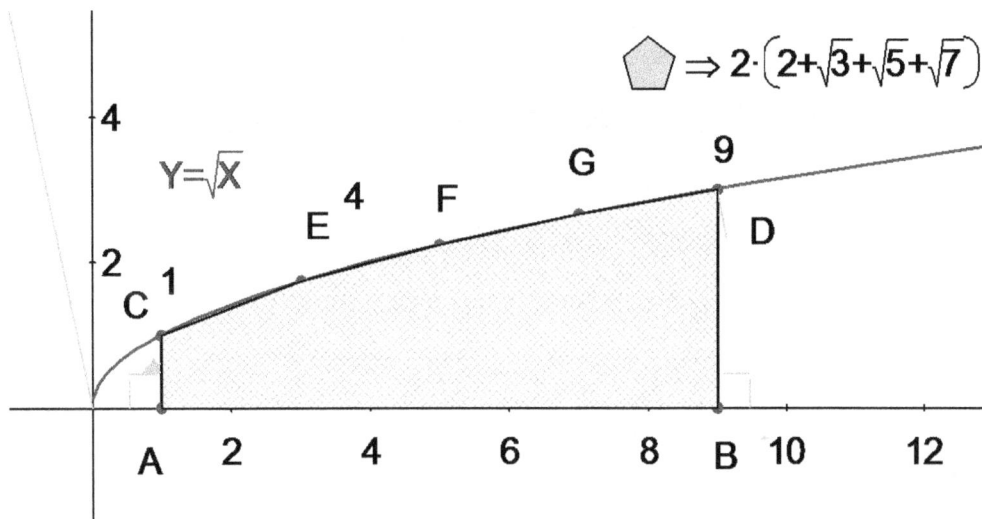

Q7. Is this an over or an under estimation? Why?

A7: The chords are under the curve, so the approximation is an under estimation. This can be confirmed by calculating the actual value of the area under the curve:

$$\int_1^9 \sqrt{x}\,dx = \frac{2x^{\frac{3}{2}}}{3}\bigg|_1^9 = \frac{2(27-1)}{3} = \frac{52}{3} < 17.2$$

Q8. Let $f(x) = 1 + \cos x$. What is the approximated value of $\int_0^\pi (1+\cos x)\,dx$ using the Trapezoidal Method with $n = 4$?

7. Show your calculations and confirm them with the help of the software.

A8:

$$\int_0^\pi (1+\cos x)\,dx \approx \frac{\pi}{2\cdot 4}\left(1+\cos 0 + 2\left(1+\cos\frac{\pi}{4}\right) + 2\left(1+\cos\frac{\pi}{2}\right) + 2\left(1+\cos\frac{3\pi}{4}\right) + 1+\cos\pi\right) = \pi$$

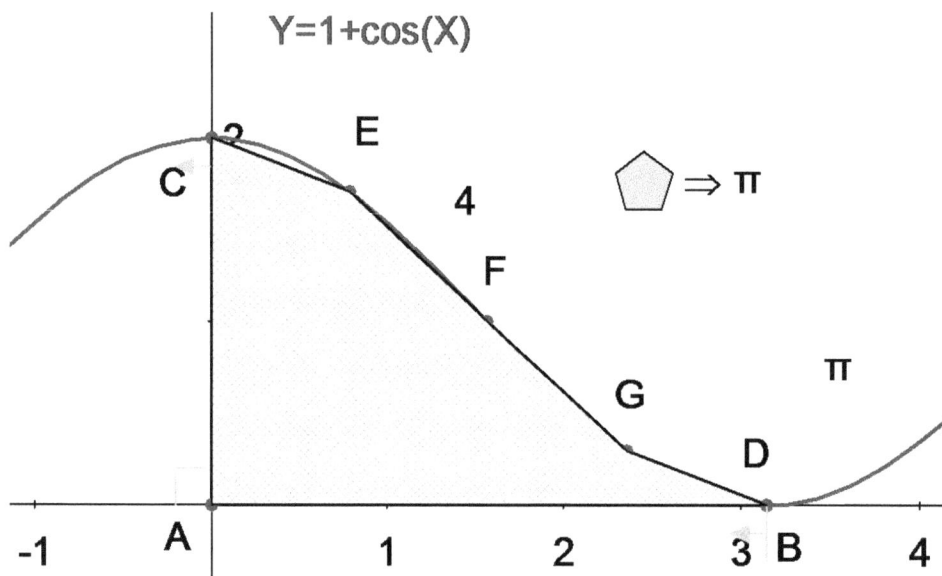

Q9. Is this an over or an under estimation? Why?

A9: *It is hard to say without evaluating an integral since some chords are above and some are below the curve on the interval* $[0, \pi]$. $\int_0^\pi (1 + \cos x)dx = x + \sin x \big|_0^\pi = \pi$ *so the approximation gave the exact value. This is due to the fact that the function is symmetrical relative to the point of inflection,* $\left(\dfrac{\pi}{2}, 0\right)$.

Q10. Based on your analysis, can you make a general statement about over and under estimations using the Trapezoidal Method?

A10: *When a function is concave down, the Trapezoidal Method gives an over-estimation, since the chords of the trapezoids are above the curve. When a function is concave up, the chords are under the curve and the Trapezoidal Method gives an under-estimation. When the function changes concavity on the interval of integration, it is not possible to state in advance. For symmetric functions it could be that the Trapezoidal Method provides the exact answer.*

Q11. How can you increase the precision of the estimation?

A11: *By increasing the number of intervals.*

The Trapezoidal Method

Exploration 4.5: Consider a plane region bounded by the graphs of equations $y = f(x)$, $y = 0$, $x = a$, and $x = b$, where $f(x)$ is continuous on $[a, b]$. You need to evaluate the area of the region. One way to approximate the area is to use n right angled trapezoids with equal heights determined by $h = \dfrac{b-a}{n}$. Your task is to develop the Trapezoidal Method for approximating the area and determine when the approximation is an over or under estimation.

TRAPEZOIDAL METHOD

1. Draw a function $y = f(x)$.

2. Approximate the graph of $y = f(x)$ from $x = a$ to $x = b$ with a polygonal chain of 4 segments. Construct a polygon containing the polygonal chain that approximates the plane region.

Q1. What is the area of the region bounded by the curve approximated with 4 line segments, $y = 0$, $x = a$, and $x = b$?

3. Verify your calculations using the software.

Q2. What is the formula for the area if you approximate the curve with a polygonal chain of n segments?

4. Verify your calculations using the software.

OVER AND UNDER

Q3. What do you need to know to determine if the estimated value of the area is larger or smaller than the actual value of the area?

Q4. Let $f(x) = x^2 + 1$. What is the approximated value of $\displaystyle\int_{-2}^{2} (x^2 + 1)dx$ using the Trapezoidal Method with $n = 4$?

5. Show your calculations and confirm them with the help of the software.

Q5. Is this an over or an under estimation? Why?

Q6. Let $f(x) = \sqrt{x}$. What is the approximated value of $\int_1^9 \sqrt{x}\,dx$ using the Trapezoidal Method with $n = 4$?

6. Show your calculations and confirm them with the help of the software.

Q7. Is this an over or an under estimation? Why?

Q8. Let $f(x) = 1 + \cos x$. What is the approximated value of $\int_0^\pi (1 + \cos x)\,dx$ using the Trapezoidal Method with $n = 4$? Show your calculations and confirm them with the help of the software.

Q9. Is this an over or an under estimation? Why?

Q10. Based on your analysis, can you make a general statement about over and under estimations using the Trapezoidal Method?

Q11. How can you increase the precision of the estimation?

4.6 Minimum Area

Exploration 4.6: Points A and B are located on the parabola $y = x^2 + 1$, such that $x_b = -x_a$. Point O is located on the parabola such that $x_b \leq x_o \leq x_a$. Find the smallest possible sum of the areas of the two parabolic segments formed by the parabola and chords OA and OB.

SUMMARY

Mathematics Objectives:

- Find an area between the joined chords of a parabola and the parabola as a function of x-coordinate of a joint point.

- Use optimization to find the position of the joint point that provides the smallest combined area of the parabolic segments.

Vocabulary:

- Parabolic segment

- Optimization

Pre-requisites:

- Derivatives of basic functions

- The 1st and 2nd derivative tests

- Area between the curves

- Definite integral

Problem Notes:

- Students first use their knowledge of definite integrals to find the area of parabolic segments formed by adjoining chords on a parabola as a function of the x-coordinate of their joint point.

- Students then use the optimization process to determine the x coordinate for the joint point on the parabola where the area of the two parabolic segments is a minimum. The software provides students with ways to visualize the problem and to confirm their calculations.

- Students should justify the minimum using either the 1st derivative or the 2nd derivative test.

<u>Technology skills:</u>

- Draw: function, line segment, arc

- Construct: polygon, tangent line

- Constrain: point proportional along the curve

- Calculate: area, slope, coordinates of a point

<u>Extensions:</u>

1. Consider the case when $x_b \neq -x_a$ and solve the same problem.

2. Given points A, P, Q, and B on the parabola $y = x^2 + 1$, such that $x_A < x_P < x_Q < x_B$, $x_A < 0$, and $x_B > 0$. Find x-coordinates of points P and Q that will minimize the sum of three parabolic segments formed by AP, PQ, and QB.

STEP-BY-STEP INSTRUCTIONS

AREA OF PARABOLIC SEGMENTS

1. Draw a function $f(x) = x^2 + 1$ and chords OA and OB.

    a. Use **Toggle grid and axes** to show the axes without the grid.

    b. Choose **Draw** → Function. Select Type → Cartesian. In the Y = prompt type x^2+1.

    c. Choose **Draw** → Line Segment and plot chords OA and OB.

    d. Select point A and the parabola, and choose **Constrain** → Point proportional. Type *a* in the open edit box.

    e. Select point B and the parabola, and choose **Constrain** → Point proportional. Type -*a* in the open edit box.

    f. Select point O and the parabola, and choose **Constrain** → Point proportional. Type *t* in the open edit box.

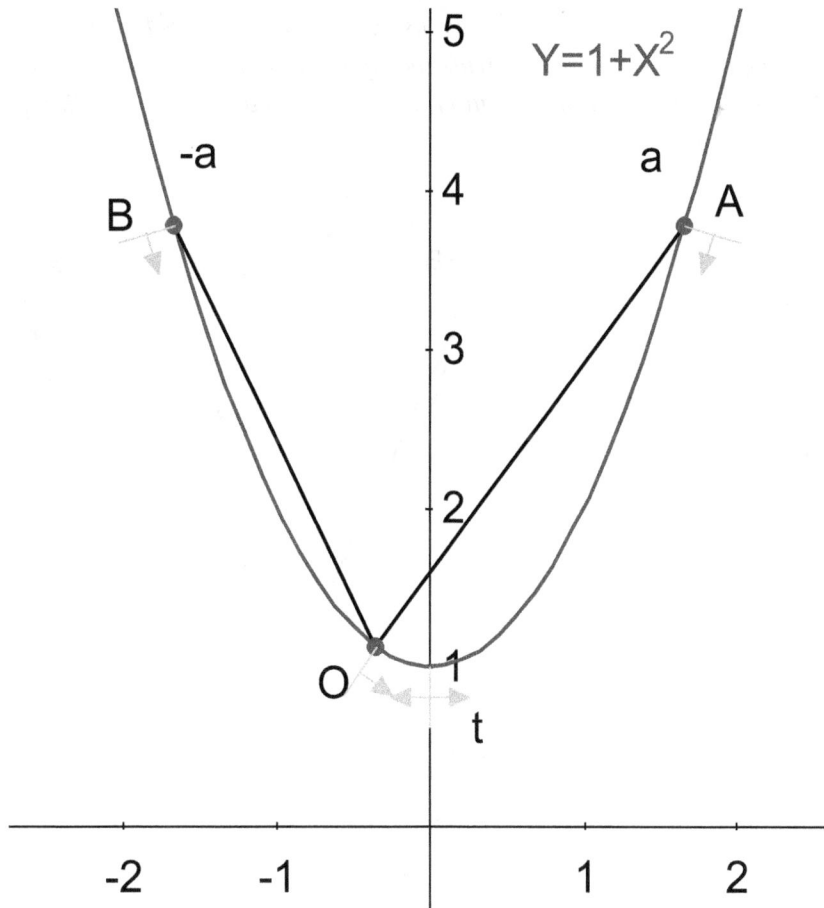

$Y=1+X^2$

Q1. Where should point O be located so that the sum of the areas of the two parabolic segments is a minimum?

A1: *Because the parabola has reflection symmetry over the y-axis, the minimum occurs when point O is at the vertex of the parabola. By dragging point O along the parabola, students should observe that the area reaches a maximum when point O coincides with either point A or point B.*

Q2. What is the area of a parabolic segment bounded by the arc and the chord OB? Setup and find the definite integral for the area of the parabolic segment.

A2: *The equation of the chord is* $y-(a^2+1)=\dfrac{a^2+1-(t^2+1)}{-a-t}(x+a) \Leftrightarrow y=(t-a)x+ta+1.$

Note: Students can verify their equations using the software. Selecting the curve and calculating an implicit equation will provide the equation of the line that contains the chord. Students will have to show that the equation they determined and the equation given by the software are equivalent.

$$\Rightarrow a+t+a^2 \cdot t+a \cdot t^2+Y \cdot (-a-t)+X \cdot \left[-a^2+t^2\right]=0$$

$$A(t) = \int_{-a}^{t} \left((t-a)x+ta+1-(x^2+1)\right)dx = \left[\frac{(t-a)x^2}{2}-\frac{x^3}{3}+tax\right]_{-a}^{t}$$

$$=\left[\frac{(t-a)t^2}{2}-\frac{t^3}{3}+t^2a\right]-\left[\frac{(t-a)a^2}{2}+\frac{a^3}{3}-ta^2\right]=\frac{(a+t)^3}{6}$$

2. Confirm your formula with the software.

    a. Choose **Draw** → Arc and plot an arc OB.

    b. Select the arc OB and chord OB, and choose **Construct** → Polygon.

    c. Select polygon interior and choose **Calculate** → Area.

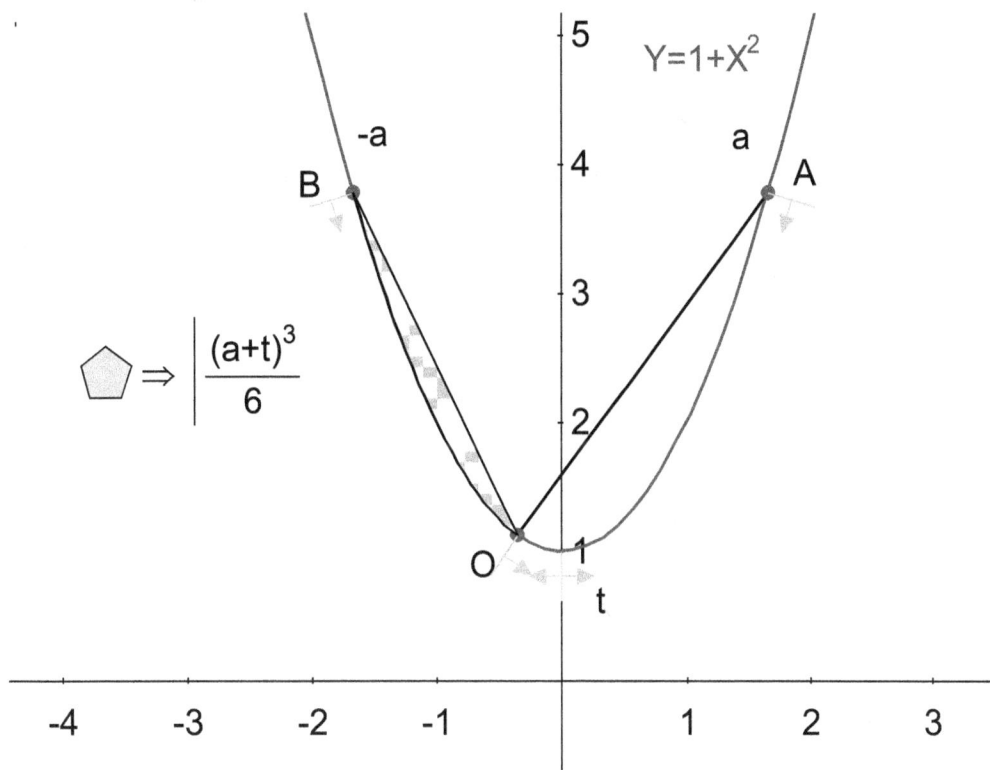

$Y=1+X^2$

$$\bigpentagon \Rightarrow \left| \frac{(a+t)^3}{6} \right|$$

Q3. What is the area of a parabolic segment bounded by the arc and the chord OA? Setup and find the definite integral for the area of the parabolic segment.

A3: *The equation of the chord is* $y-(a^2+1)=\dfrac{a^2+1-(t^2+1)}{a-t}(x-a) \Leftrightarrow y=(t+a)x-ta+1.$

$$A(t)=\int_t^a \left((t+a)x-ta+1-(x^2+1)\right)dx = \left[\frac{(t+a)x^2}{2}-\frac{x^3}{3}-tax\right]_t^a$$

$$=\left[\frac{(t+a)a^2}{2}-\frac{a^3}{3}-ta^2\right]-\left[\frac{(t+a)t^2}{2}-\frac{t^3}{3}-t^2a\right]=\frac{(a-t)^3}{6}$$

3. Confirm your formula with the software.

   a. Choose **Draw** → Arc and plot an arc OA.

   b. Select the arc OA and chord OA, and choose **Construct** → Polygon.

   c. Select polygon interior and choose **Calculate** → Area.

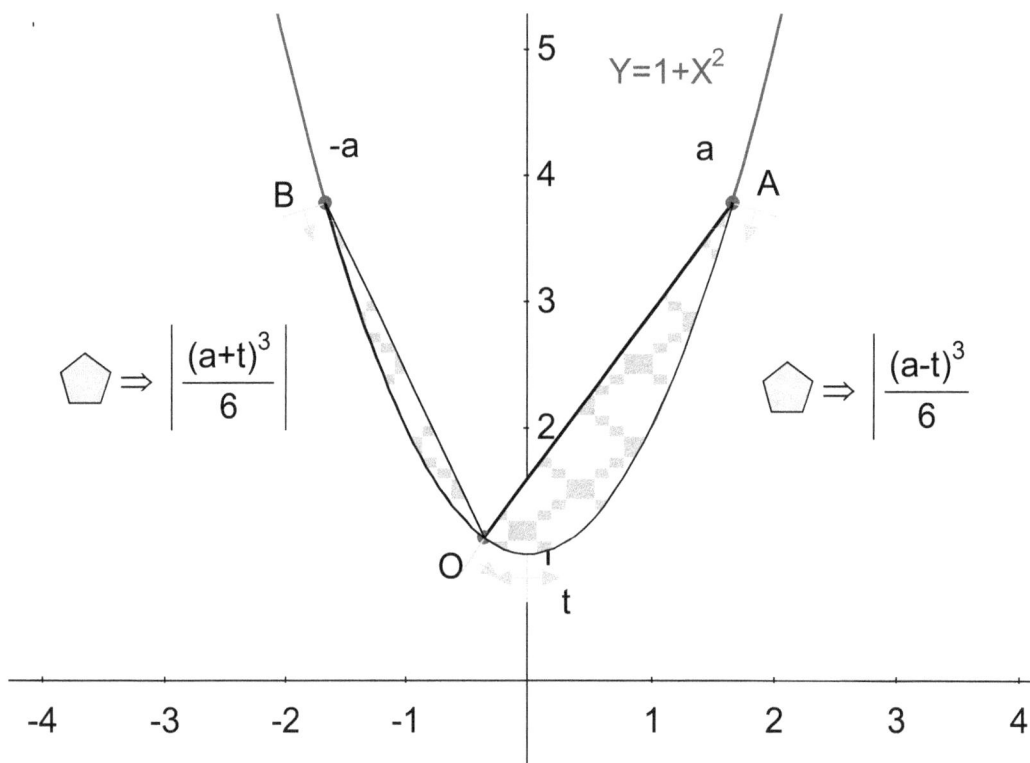

Q4. Determine the value of t such that the sum of the areas of the two parabolic segments is a minimum value. What is that value of t?

A4: $A(t) = \dfrac{(a+t)^3 + (a-t)^3}{6} = \dfrac{a^3 + 3at^2}{3}$, $\dfrac{dA}{dt} = 2at = 0$, when $t = 0$. $A(0) = \dfrac{1}{3}a^3$.

4. Verify your calculations using the software.

   a. Open a new document.

   b. Choose **Draw** → Function. Select Type → Cartesian. In the Y = prompt type area function in terms of x.

   c. Select the plot of the function and choose **Construct** → Tangent .

   d. Select the point of tangency and the curve and choose **Constrain** → Point proportional. In the open edit box type *0*.

   e. Select the tangent line and choose **Calculate** → Slope. Select the point of tangency and choose **Calculate** → Coordinates.

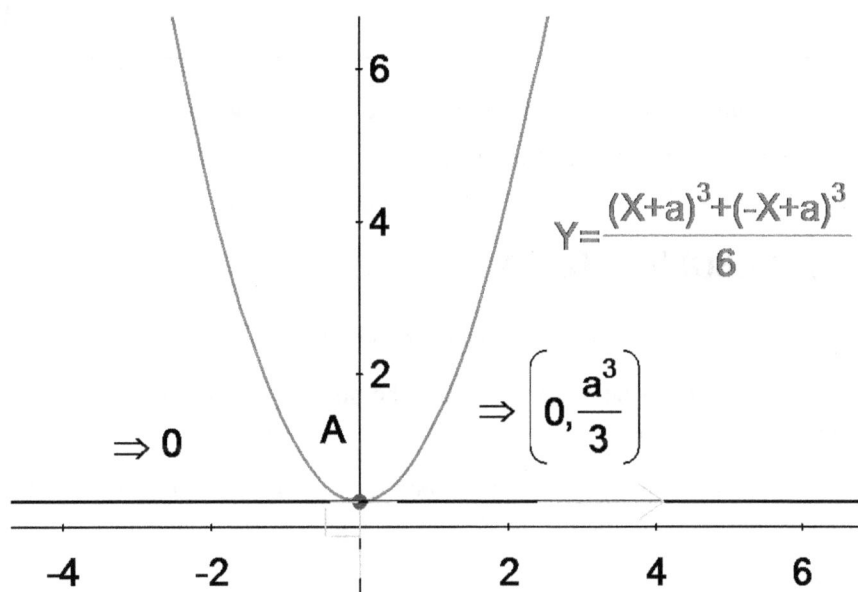

$$Y = \frac{(X+a)^3 + (-X+a)^3}{6}$$

$$\Rightarrow \left(0, \frac{a^3}{3}\right)$$

$\Rightarrow 0$

A

Note: Students should justify that t=0 is a minimum using either the 1ˢᵗ derivative or 2ⁿᵈ derivative test.

Exploration 4.6: Points A and B are located on the parabola $y = x^2 + 1$, so that $x_b = -x_a$. Point O is located on the parabola so that $x_b \leq x_o \leq x_a$. Find the smallest possible sum of the areas of the two parabolic segments formed by the parabola and chords OA and OB.

AREA OF PARABOLIC SEGMENTS

1.   Draw a function $f(x) = x^2 + 1$ and chords OA and OB.

Q1. Where should point O be located so that the sum of the areas of the two parabolic segments is a minimum?

Q2. What is the area of a parabolic segment bounded by the arc and the chord OB? Setup and find the definite integral for the area of the parabolic segment.

2.   Confirm your formula with the software.

Q3. What is the area of a parabolic segment bounded by the arc and the chord OA? Setup and find the definite integral for the area of the parabolic segment

3.   Confirm your formula with the software.

Q4. Determine the value of t such that the sum of the areas of the two parabolic segments is a minimum value. What is that value of t?

4.   Verify your calculations using the software.

5. Differential Equations

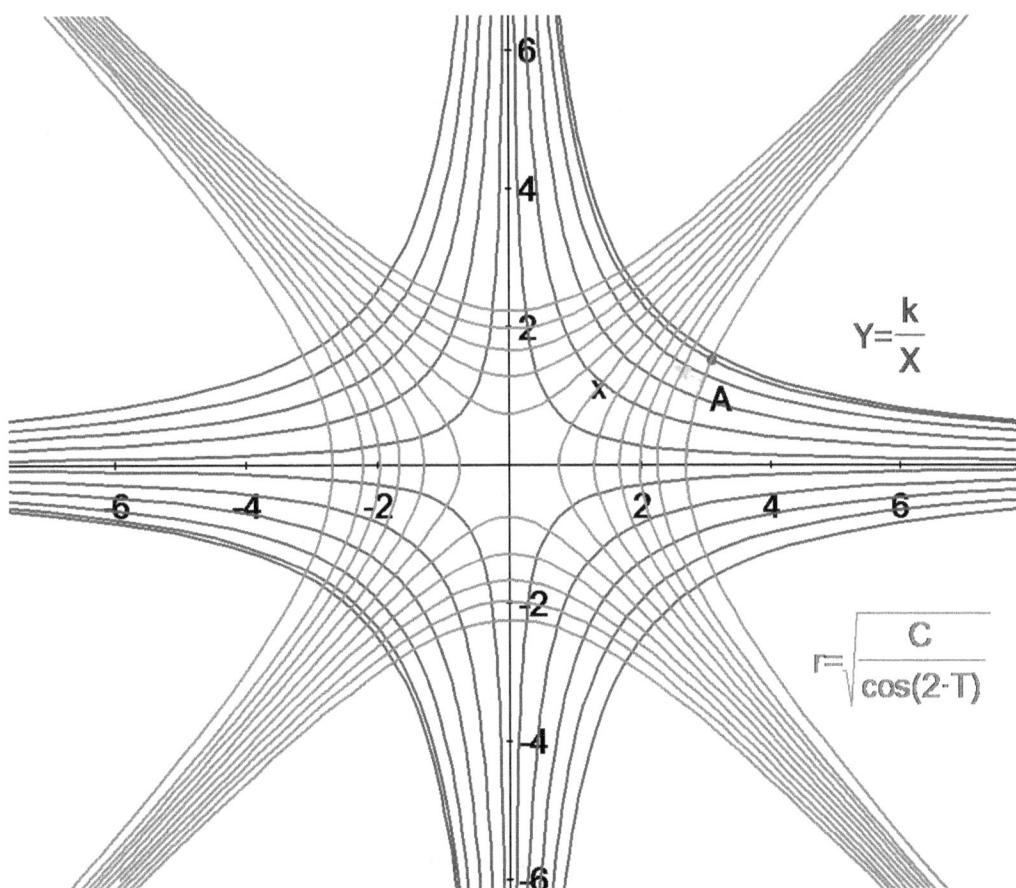

$$Y = \frac{k}{X}$$

$$r = \sqrt{\frac{C}{\cos(2 \cdot T)}}$$

5.1 Orthogonal Trajectory to a Circle

<u>Exploration 5.1:</u> Two families of curves are said to be *mutually orthogonal* when each curve in one of the families is orthogonal to all members of another family. Then each curve in one of the families is called an *orthogonal trajectory* of the other family. In this problem you will explore orthogonal trajectories for the family of circles, $x^2 + y^2 = r^2$, where r is the radius of the circle.

SUMMARY

<u>Mathematics Objectives:</u>

- Derive the differential equation for the orthogonal trajectories to a family of circles

- Find the general solution of the differential equation

- Explore families of orthogonal trajectories graphically and confirm their mutual orthogonality using the software

<u>Vocabulary:</u>

- Orthogonal trajectory

- General and particular solutions of differential equations

<u>Pre-requisites:</u>

- Derivative as a slope of a tangent line

- The angle between curves

- The slopes of perpendicular lines.

- Separation of variables for solving 1st order linear differential equations.

<u>Problem Notes:</u>

- The question of orthogonal trajectories is usually considered as an application problem for solving 1st order linear differential equations. However, the concept of an angle between the curves and how to find orthogonal trajectories is important on its own in both mathematics and physics. For example, in electrostatics the circles could represent equipotential lines and orthogonal trajectories would represent the electric field lines.

- The angle between the curves is defined as the angle between tangent lines to each curve at the point of intersection. If students have not been introduced to this idea, this problem provides an excellent opportunity to introduce this definition.

- Students first explore the slope of the tangent line to the graph of $x^2 + y^2 = r^2$ and derive a differential equation for the orthogonal curve using the fact that tangent lines of orthogonal curves are perpendicular.

- Students then solve the differential equation by separation of variables and plot the general solution of a differential equation. They verify that these families of curves are mutually orthogonal with the help of the software.

Technology skills:

- Draw: circle, line segment, function
- Constrain: radius, perpendicular, point proportional along the curve
- Construct: tangent line, trace
- Calculate: slope, angle

Extension:

Given two families of circles defined by the equations $(x-a)^2 + y^2 = a^2 + c$ and $x^2 + (y-b)^2 = b^2 - c$, where a, b, and c are real. Prove that these two families of curves are mutually orthogonal.

STEP-BY-STEP INSTRUCTIONS

STARTING POINT

Q1. How can we determine an angle between two intersecting curves?

A1: *The angle between the curves is defined as the angle between tangent lines to each curve at the point of intersection.*

Note: if students have not been introduced the idea of an angle between the curves, the teacher can use this opportunity to introduce this definition.

EXPLORING SLOPES OF TANGENT LINES

Q2. Make a prediction: What functions are orthogonal trajectories of the family of circles, $x^2 + y^2 = r^2$, where r is the radius of the circle?

A2:   Since all diameters are perpendicular to the tangent lines of a circle, orthogonal trajectories are lines through the origin.

1. To help visualize Q2, draw a circle with the center at the origin and constrain its radius to be r.

    a. Use **Toggle grid and axes** to show the axes without a grid.

    b. Choose **Draw** → Point and plot point A at the origin

    c. Choose **Draw** → Circle. Select the circle and choose **Constrain** → Radius. In the open edit box type r. Hide point B.

2. Draw a tangent line to the circle at an arbitrary point C in the 1st quadrant.

    a. Select the circle and choose **Construct** → Tangent to Curve. The tangent line and the point of tangency will be displayed (point C).

3. Drop a perpendicular from point C to a point D on the x-axis. Constrain the x-coordinate of point D to be a.

    a. Choose **Draw** → Segment and construct a segment from point C to the x-axis. Select this segment and the x-axis and choose **Constrain** → Perpendicular.

    b. Select point D and the x-axis and choose **Constrain** → Point proportional. Type a in the edit box.

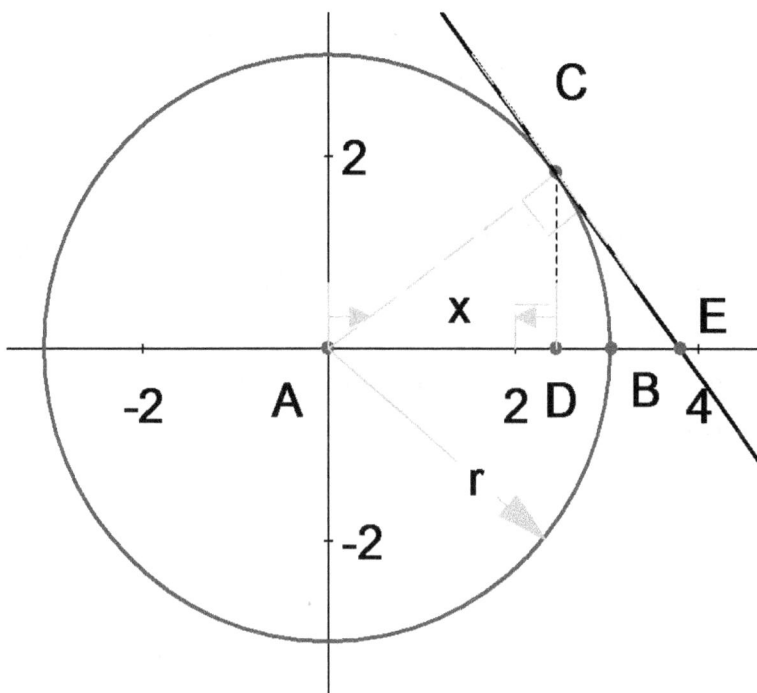

Q3. What is the slope of the tangent line at point C?

A3: $\dfrac{-a}{\sqrt{r^2-a^2}}$. *If using calculus, students could take the derivative of the function*

$y=\sqrt{r^2-x^2}$. *From geometric considerations,* $\angle ACD = \angle CED$ *so*

$\tan(\angle CED) = \dfrac{CD}{ED} = \dfrac{AD}{CD} = \dfrac{x}{y}\Big|_{x=a} = \dfrac{a}{\sqrt{r^2-a^2}}$. *The slope of the tangent line*

$m = \tan\left(\pi - \angle CED\right) = -\tan(\angle CED) = -\dfrac{a}{\sqrt{r^2-a^2}}$

4. Verify your answer using *Geometry Expressions.*

   a. Select the tangent line and choose **Calculate → Slope**.

Q4. What is the slope of the tangent line for the curve that is orthogonal to the circle at point C?

A4:   Since orthogonal curves are defined as curves with perpendicular tangent lines at points of intersection, the slope of the tangent line of the orthogonal trajectory is the negative reciprocal of the slope of the tangent line to the circle: $\dfrac{\sqrt{r^2 - a^2}}{a}$.

SOLVING A DIFFERENTIAL EQUATION

Q5.   Setup and solve a differential equation for the orthogonal curve. Use the condition that the curve contains point C.

A5:   The slope of the function at x is a derivative of the function. Thus, the differential equation can be written as $\dfrac{dy}{dx} = \dfrac{y}{x}$ *with initial condition* $y(a) = \sqrt{r^2 - a^2}$.

Solving this equation by separation of variables (considering points in quadrant I for simplicity): $\dfrac{dy}{y} = \dfrac{dx}{x} \Leftrightarrow \ln y = \ln x + \ln C \Leftrightarrow y = Cx$. *This is a family of lines through the origin that represents orthogonal trajectories to a family of circles. The solution holds true for all points on the circle, except x=0 and y=0. Thus, the coordinate axes, y = 0 and x = 0, need to be added to the family of orthogonal trajectories. In order to find a specific function that passes through the point C substitute the coordinates of point C to find the constant of integration:* $\sqrt{r^2 - a^2} = Ca \Leftrightarrow C = \dfrac{\sqrt{r^2 - a^2}}{a}$. *The final equation for this particular solution is* $y = \dfrac{\sqrt{r^2 - a^2}}{a} x$.

5.   Graph this function, and drag point D to observe the angle between the two curves.

   a.   Choose **Draw** → Function. Select Type → Cartesian. In the Y = prompt type the expression for the function.

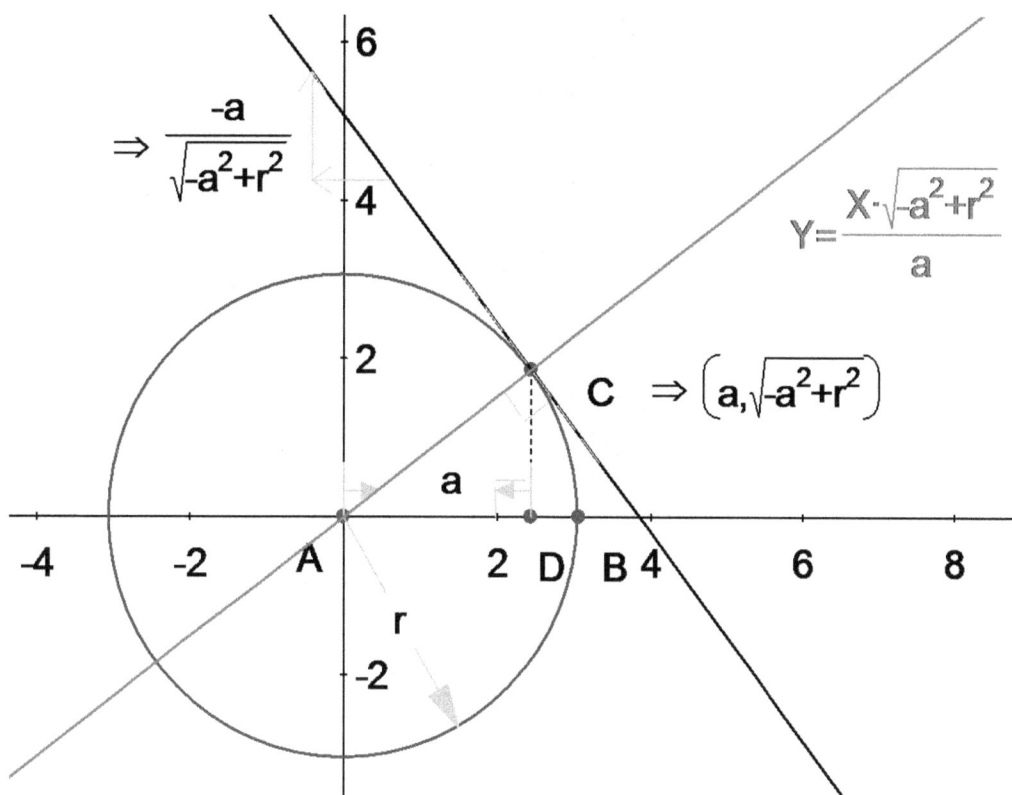

$$\Rightarrow \frac{-a}{\sqrt{-a^2+r^2}}$$

$$Y = \frac{X \cdot \sqrt{-a^2+r^2}}{a}$$

$$C \Rightarrow \left(a, \sqrt{-a^2+r^2}\right)$$

Q6. What is the angle between the circle and the curve you found?

A6: The angle is 90°.

6. Using *Gx's* **Trace** tool, draw both families of curves. You may want to remove the tangent line to the circle and points C and D to unclutter the picture.

   a. Select the tangent line to the circle and point of tangency and click delete.

   b. Click on the trajectory you drew and choose **Trace.** In the open window choose the parameter *a*; set the start, end and count values according to the number of lines you want to display. Next, trace the family of circles by selecting the circle and choosing the parameter *r* in the trace window. Set the start and end and count values according to the number of circles you want to display.

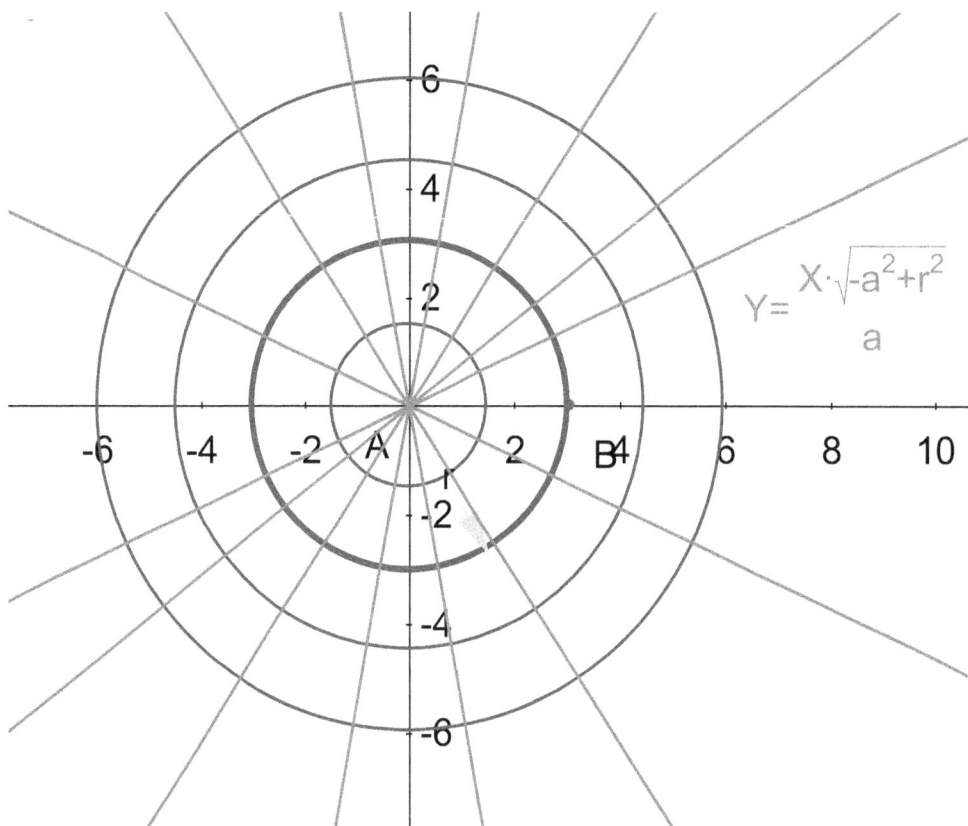

$$Y= \frac{X \cdot \sqrt{-a^2 + r^2}}{a}$$

Orthogonal Trajectory to a Circles

Exploration 5.1: Two families of curves are said to be *mutually orthogonal* when each curve in one of the families is orthogonal to all members of another family. Then each curve in one of the families is called an *orthogonal trajectory* of the other family. In this problem you will explore orthogonal trajectories for the family of circles, $x^2 + y^2 = r^2$, where r is the radius of the circle.

STARTING POINT

Q1. How can we determine an angle defined between two intersecting curves?

EXPLORING SLOPES OF TANGENT LINES

Q2. Make a prediction: What functions are orthogonal trajectories of the family of circles, $x^2 + y^2 = r^2$, where r is the radius of the circle?

1. To help visualize Q2, draw a circle with the center at the origin and constrain its radius to be r.

2. Draw a tangent line to the circle at an arbitrary point C in the 1st quadrant.

3. Drop a perpendicular from the point C to a point D on the x-axis. Constrain the x-coordinate of the point D to be a.

Q3. What is the slope of the tangent line at point C?

4. Verify your answer using *Geometry Expressions*.

Q4. What is the slope of the tangent line for the curve that is orthogonal to the circle at point C?

SOLVING A DIFFERENTIAL EQUATION

Q5. Setup and solve a differential equation for the orthogonal curve. Use the condition that the curve contains point C.

5. Graph this function, and drag point D to observe the angle between the two curves.

Q6. What is the angle between the circle and the curve you found?

6. Using *Gx's* **Trace** tool to draw both families of curves. You may want to remove the tangent line to the circle and points C and D to unclutter the picture.

5.2 Orthogonal Trajectory to a Hyperbola

<u>Exploration 5.2:</u> Two families of curves are said to be *mutually orthogonal* when each curve in one of the families is orthogonal to all members of another family. Then each curve in one of the families is called an *orthogonal trajectory* of the other family. In this problem you will explore orthogonal trajectories for the family of hyperbolas, $y = \dfrac{k}{x}$.

SUMMARY

<u>Mathematics Objectives:</u>

- Derive the differential equation for the orthogonal trajectories to a family of hyperbolas

- Find the general solution of the differential equation

- Explore families of orthogonal trajectories graphically and confirm their mutual orthogonality using the software

<u>Vocabulary:</u>

- Orthogonal trajectory

- General and particular solutions of differential equations

<u>Pre-requisites:</u>

- Derivative as a slope of the tangent line

- The angle between curves

- The slopes of perpendicular lines.

- Separation of variables for solving 1st order linear differential equations.

<u>Problem Notes:</u>

- The question of orthogonal trajectories is usually considered as an application problem for solving 1st order linear differential equations. However, the concept of an angle between the curves and how to find orthogonal trajectories is important on its own in both mathematics and physics. For example, in engineering analysis in heat transfer problems, the lines of heat flux are perpendicular to the lines of constant temperature (isotherms).

- The angle between the curves is defined as the angle between tangent lines to each curve at the point of intersection. If students have not been introduced to this idea, this problem provides an excellent opportunity to introduce this definition.

- Students first explore the slope of the tangent line to the graph of $y = \dfrac{k}{x}$ and derive a differential equation for the orthogonal curve using the fact that tangent lines of orthogonal curves are perpendicular.

- Students then solve the differential equation by separation of variables and plot the general solution of a differential equation. They verify that these families of curves are mutually orthogonal with the help of the software.

Technology skills:

- Draw: function

- Constrain: point proportional along the curve

- Construct: tangent line, trace

- Calculate: slope, angle

Extension:

Given two families of parabolas defined by the equations $4a(a-y) = x^2$ and $4b(b+y) = x^2$, where $a > 0$ and $b > 0$. Prove that these two families of curves are mutually orthogonal.

STEP-BY-STEP INSTRUCTIONS

STARTING POINT

Q1. How can we determine an angle between two intersecting curves?

A1: The angle between the curves is determined as the angle between tangent lines to each curve at the point of intersection.

Note: if students have not been introduced to the idea of the angle between the curves, the teacher can use this opportunity to introduce this definition.

EXPLORING SLOPES OF TANGENT LINES

Q2. Can you describe an orthogonal trajectory for the family of hyperbolas $y = \dfrac{k}{x}$?

A2: The answers will vary. Suggest students try to sketch a slope field for a family of hyperbolas and try to imagine the shape of their orthogonal trajectories.

1. Draw a hyperbola $y = \dfrac{k}{x}$.

    a. Use **Toggle grid and axes** to show the axes without a grid.

    b. Choose **Draw** → Function. Choose Cartesian for the Type, and in the Y= type the expression for the function.

2. Draw a tangent line to the hyperbola at an arbitrary point $A\left(x, \dfrac{k}{x} \right)$.

    a. Select the hyperbola and choose **Construct** → Tangent. The tangent line and point of tangency will be displayed (point A).

    b. Select point A and the hyperbola, and choose **Constrain** → Point proportional. In the open edit box type *x*.

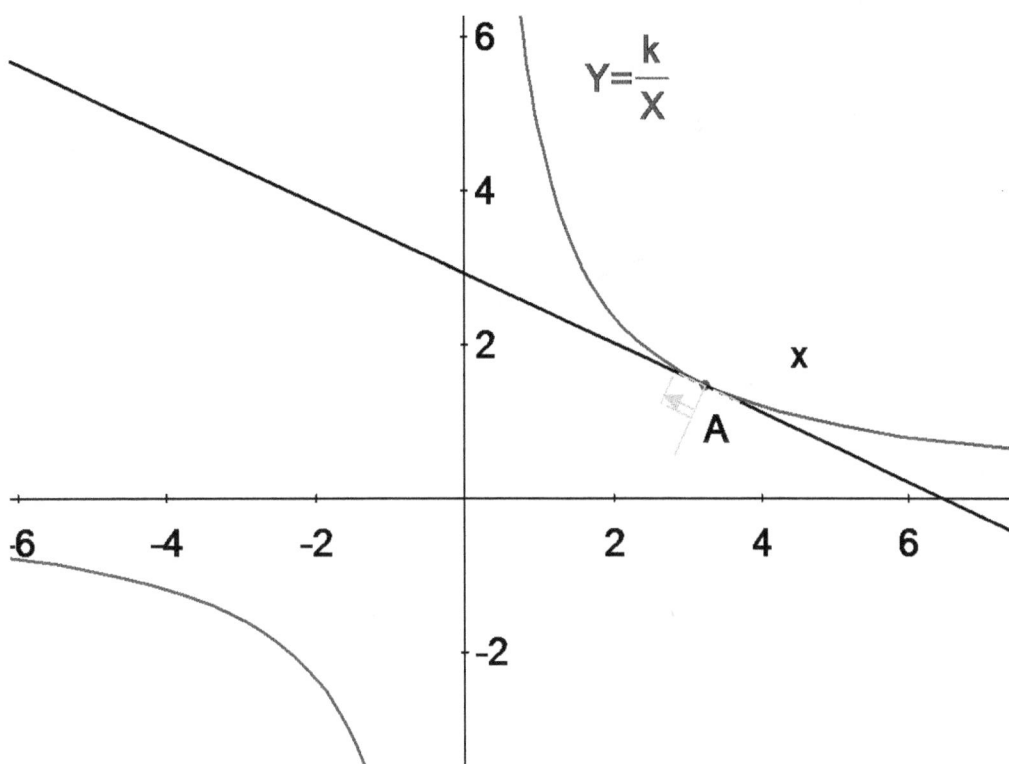

Q3.   What is the slope of the tangent line at point A?

A3:   Differentiate the equation $y = \dfrac{k}{x}$. The derivative is $y' = -\dfrac{k}{x^2}$.

4.   Verify your answer using the software.

  a.   Select the tangent line and choose **Calculate** → Slope.

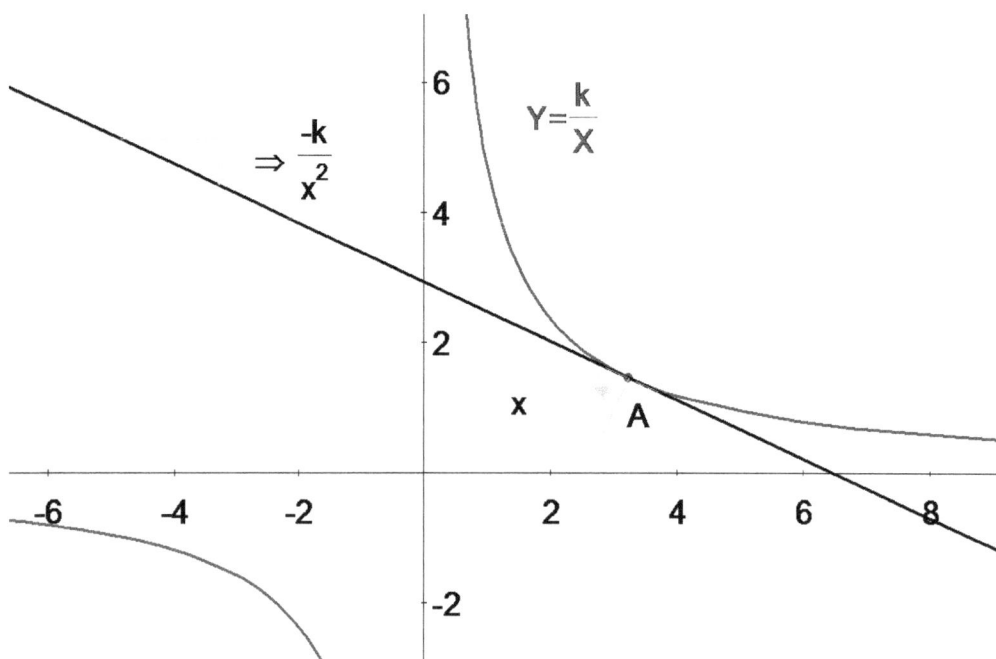

Q4. Both the derivative and the function have the constant k in their expressions. Can you express the derivative in terms of x and y only?

A4. Solve for the parameter k and substitute into the derivative to get:

$$k = xy \Rightarrow y' = -\frac{xy}{x^2} = -\frac{y}{x}$$

Q5. What is the slope of the tangent line for the curve that is orthogonal to the hyperbola at point A?

A5: $m = \dfrac{x}{y}$.

SOLVING A DIFFERENTIAL EQUATION

Q6. Setup and solve a differential equation for an orthogonal curve.

A6: $\dfrac{dy}{dx} = \dfrac{x}{y}$. *Using separation of variables:* $ydy = xdx \Leftrightarrow \dfrac{y^2}{2} = \dfrac{x^2}{2} + \dfrac{C}{2}$ *Thus, the orthogonal trajectories are hyperbolas:* $y^2 - x^2 = C$.

Note: based on this equation, the branches of hyperbolas are open up and down, and asymptotes are $y = \pm x$

5. Graph a particular orthogonal curve that contains the point A. Drag point A along the original hyperbola and observe the angle between the two curves.

Note: students can solve for y and graph both functions, $y = \pm\sqrt{x^2 + C}$ *, or they can re-write this equation in polar coordinates:* $r = \sqrt{\dfrac{C}{\sin^2\theta - \cos^2\theta}} = \sqrt{-C\sec(2\theta)}$ *.*

a. Choose **Draw** → Function. Choose appropriate Type and enter the expression for the function.

b. Select parameter C in the **Variables** menu and use the slider to adjust the value of C, so that the curve passes through point A.

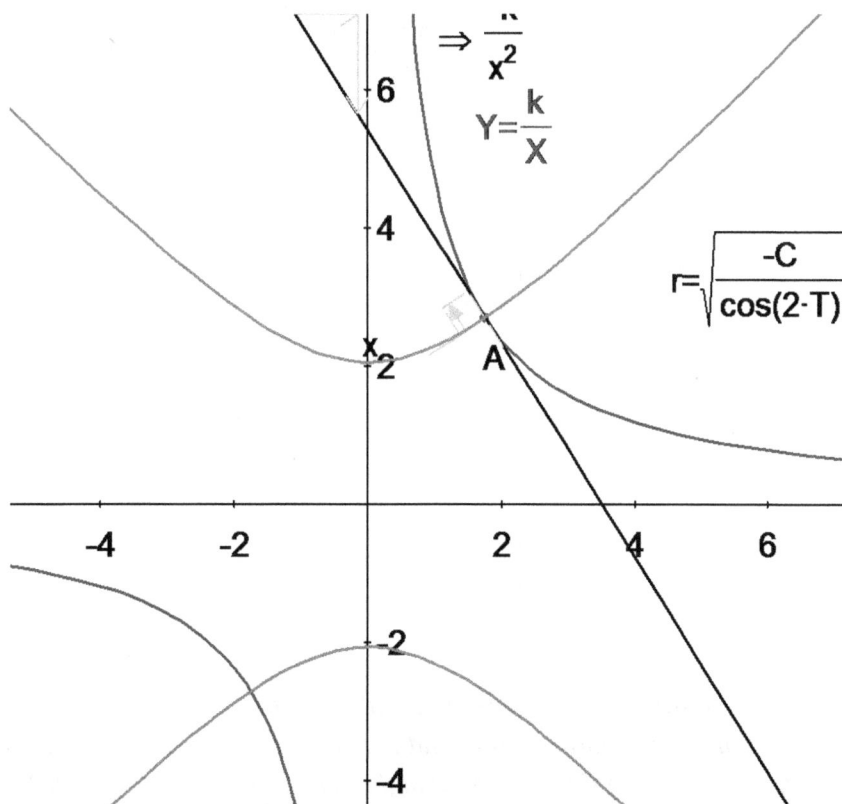

Q7. What is the angle between the hyperbola and the curve you found?

A7: The angle is 90°.

6. Verify your answer using the software.

    a. Select the new hyperbola and choose **Construct** → Tangent to Curve.

    b. Drag point B to coincide with point A. When B coincides with A select both the tangent lines and choose **Calculate** → Real Measurement → Angle.

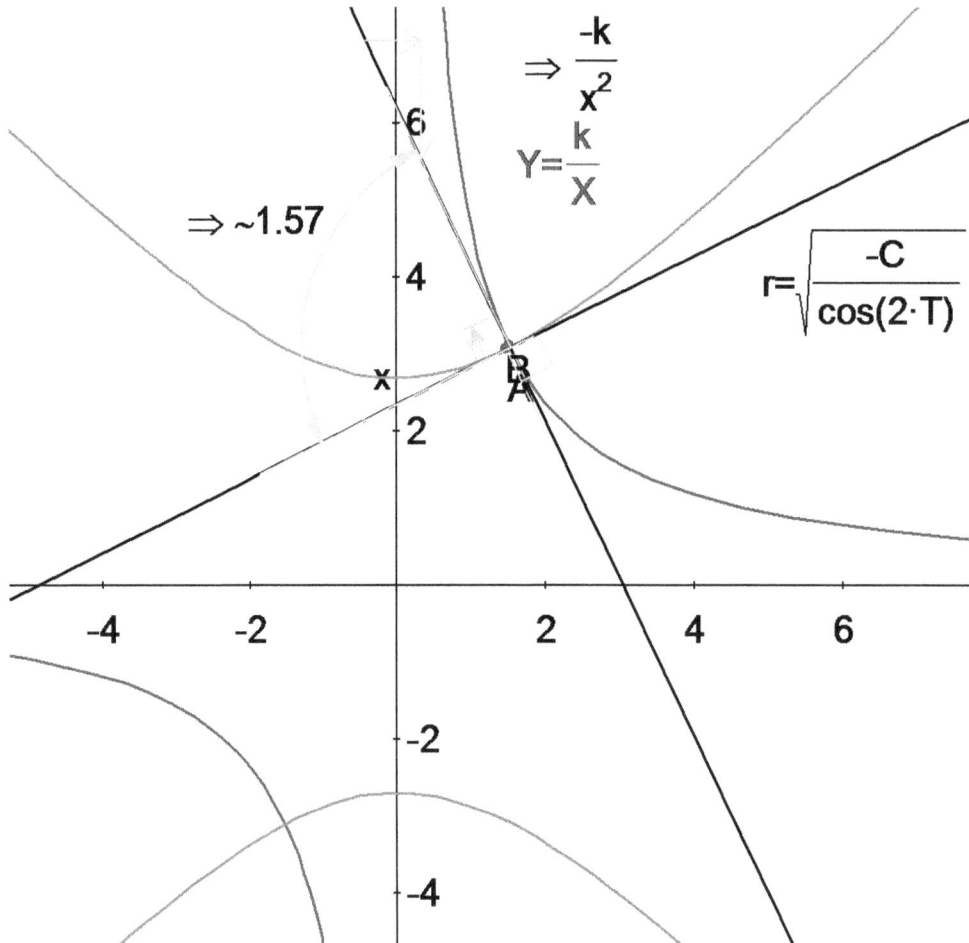

Note: the software provides an approximate value of the angle, 1.57 radians ≈ π/2. However, this is a numeric approximation and is not an exact answer. Students can move point A and adjust the value of C to observe that the curves are orthogonal. The fact that the

angle between the tangent lines will always be π/2 is based on the way the differential equation was setup.

7. Draw both families of the curves, using Trace option of the software. (Remove tangent lines and points of tangency to unclutter the picture). Confirm visually that these are orthogonal curves.

   a. Select both tangent lines and the points of tangency and click delete.

   b. Click on one of the curves and choose **Trace.** In the open window choose the correct parameter for the curve (*k* or *C*), choose start -5, end 5, and count according to the number of curves you want to be displayed.

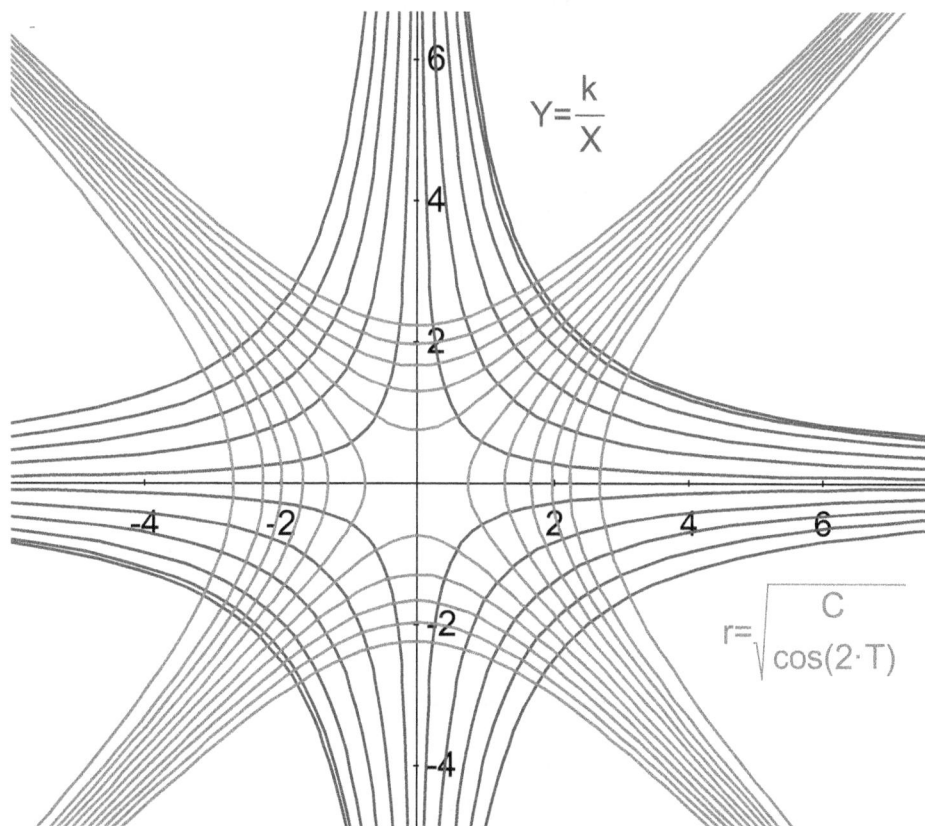

$$Y=\frac{k}{X}$$

$$r=\sqrt{\frac{C}{\cos(2\cdot T)}}$$

Orthogonal Trajectory to a Hyperbola

<u>Exploration 5.2:</u> Two families of curves are said to be *mutually orthogonal* when each curve in one of the families is orthogonal to all members of another family. Then each curve in one of the families is called an *orthogonal trajectory* of the other family. In this problem you will explore orthogonal trajectories for the family of hyperbolas, $y = \dfrac{k}{x}$.

STARTING POINT

Q1. How can we determine an angle between two intersecting curves?

EXPLORING SLOPES OF TANGENT LINES

Q2. Can you describe an orthogonal trajectory for the family of hyperbolas?

Q3. What is the slope of the tangent line at point A?

4. Verify your answer using *Geometry Expressions*.

Q4. Both the derivative and the function have the constant k in their expressions. Can you express the derivative in terms of x and y *only*?

Q5. What is the slope of the tangent line for the curve that is orthogonal to the hyperbola at point A?

SOLVING A DIFFERENTIAL EQUATION

Q6. Setup and solve the differential equation for the orthogonal curve.

5. Graph a particular orthogonal curve that contains the point A. Drag point A along the original hyperbola and observe the angle between the two curves.

Q7. What is the angle between the hyperbola and the curve you found?

6. Verify your answer using *Geometry Expressions*.

7. Draw both families of the curves, using *Gx's* **Trace** tool. (Remove tangent lines and points of tangency to unclutter the picture).

6.Sequences and Series

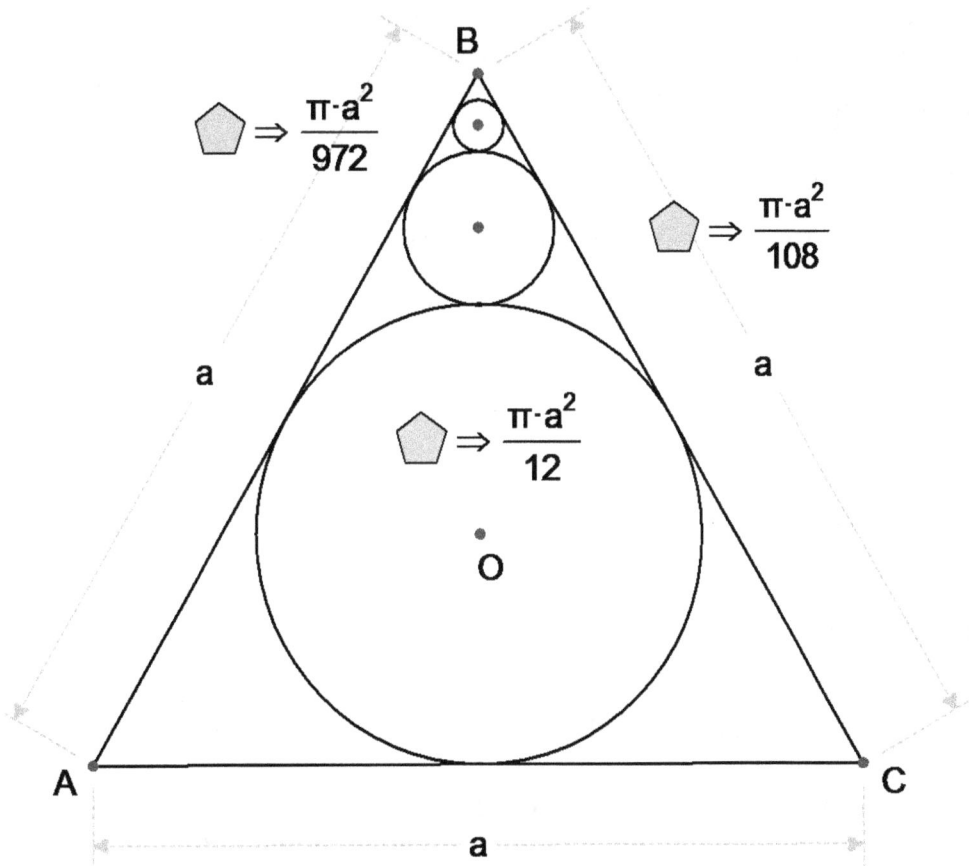

$$\pentagon \Rightarrow \frac{\pi \cdot a^2}{972}$$

$$\pentagon \Rightarrow \frac{\pi \cdot a^2}{108}$$

$$\pentagon \Rightarrow \frac{\pi \cdot a^2}{12}$$

6.1 Infinite Stairs

Exploration 6.1: A staircase leads from point A to point C and has two steps, each one is 1 foot high and 1 foot wide. You decide to create a new staircase that will still lead from point A to point C, but with 4 steps so that each step is now 0.5 feet high and 0.5 feet wide. You doubled the number of steps while keeping their heights equal to their widths. You are interested in the total length of the stairs defined as the total sum of the heights and widths of all the steps. If you continue this process, what is the total length of the stairs as the number of steps increases without bound?

SUMMARY

Mathematics Objectives:

- Find the nth term of a sequence and the partial sum of a series.

- Investigate convergence of an infinite series

Vocabulary:

- Bounded sequence

- Convergence

Pre-requisites:

- Limits at infinity

Problem Notes:

- This well-known problem challenges students' intuition. From geometry students may be more familiar with the convergence of regular polygons inscribed in a circle. The area of the polygon converges to the area of the circle, and the polygon perimeter converges to the circumference of the circle making the infinite stair problem counterintuitive. Visually, the steps are disappearing and the staircase becomes a sloped hill; thus, many students naturally predict that the length of the staircase becomes equal to the distance between the two points, A and C. However, careful consideration of the infinite series reveals that the total sum of all widths and heights of the steps remains the same.

- Students first explore the stairs with 2 steps and determine the total length of the stairs. Then they double the number of steps, by replacing each step with 2 smaller equal steps, and so on. Students analyze the pattern and determine the height and width of each step for the case of the *n*-step stairs.

- Students then determine the general equation for the length of the stairs with *n* steps and find out that it stays the same, as the number of steps increases without bound.

Technology skills:

- Draw: polygon, line segment
- Constrain: perpendicular, distance, slope
- Calculate: perimeter, area
- Construct: midpoint, translation
- Using a slider

Extension:

Given circle of radius *r*. Construct a circle with radius *r*/2, *r*/4, *r*/8, etc. as shown on the diagram below. Find the sum of the circumferences of all circles.

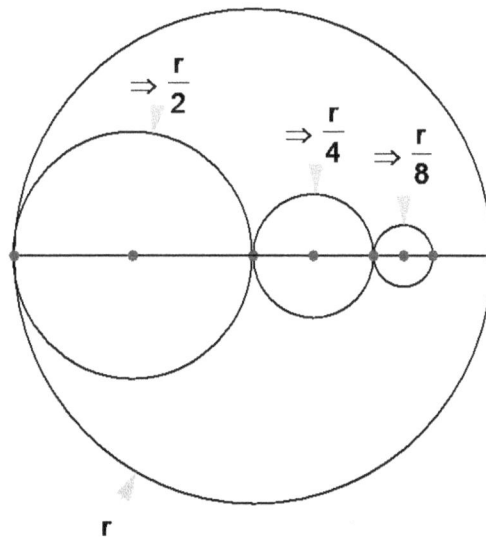

STEP-BY-STEP INSTRUCTIONS

UP THE STAIRCASE

Q1. What do you think the length of the stairs will be when the number of steps increases without bound?

A1: Answers will vary. Some students will state that the length is the hypotenuse of the triangle that is formed as a result of this process, $L = 2\sqrt{2}$ ft. Some students may predict that the length of stairs does not change in this process, and it is $L = 4$ ft.

1. Draw a staircase with two steps as shown in the diagram below.

   a. Use **Toggle grid and axes** to hide the axes and the grid.

   b. Choose **Draw** → Polygon and plot a L-shaped polygon that will represent the staircase.

   c. Select each two adjoined segments on the staircase and choose **Constrain** → Perpendicular.

   d. For each step, select the height of the step and choose **Constrain** → Distance/Length, and type 1.

   e. Repeat the Constrain Distance/Length process so that the width of each step is also 1.

   f. Select the base of the staircase and choose **Constrain** → Slope. Type 0.

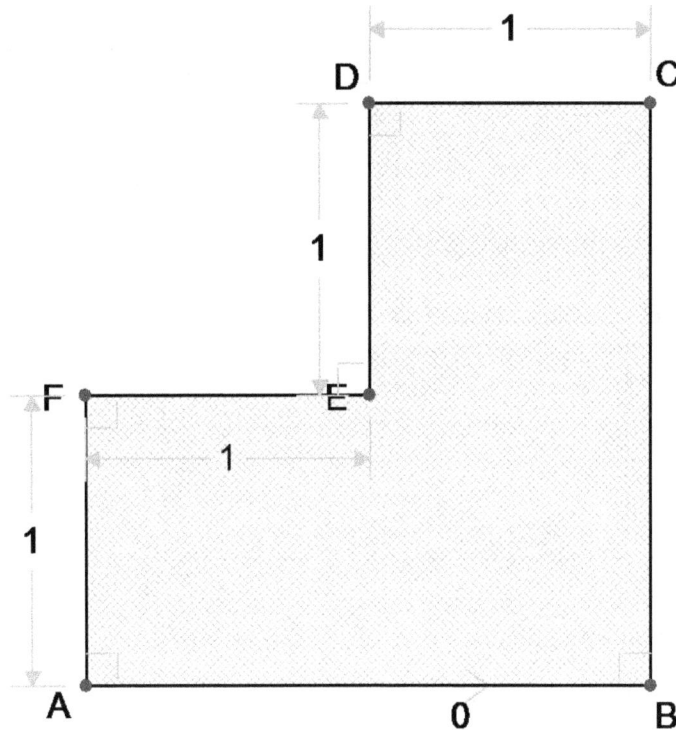

Q2. What is the total height and the total width of the stairs?

A2: The total height is 2 ft and the total width is 2 ft.

Q3. Based on your answer in question 2, what is the total length of the stairs, measured as the sum of heights and widths of the steps of the stairs?

A3: AF + FE + ED + DC = 4 ft.

2.   Verify your answer using the software.

    a.   Select the polygon and choose **Calculate** → Perimeter.

Note: Students can subtract the total height and total width from the perimeter in order to find the total length of the stairs. Since the total height and the total length of the staircase is constant, the perimeter of the polygon is different from the length of the stairs by a constant value. Thus, the value of the perimeter is sufficient for the analysis of the series.

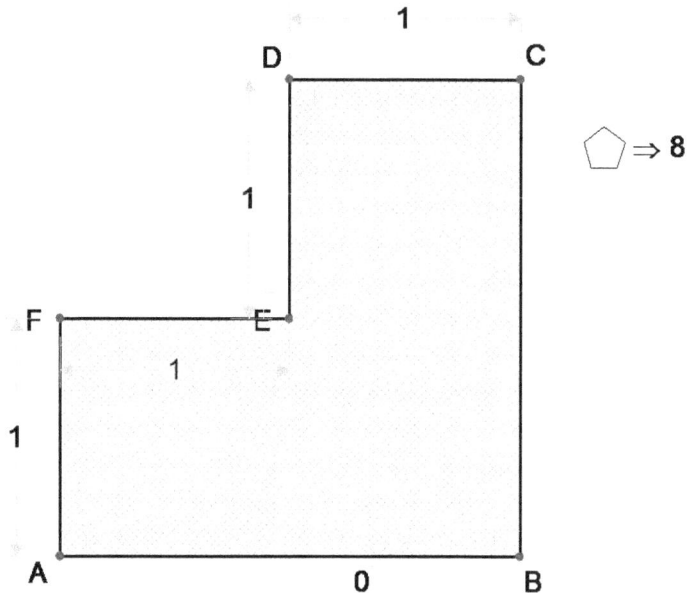

3. Draw a staircase with 4 steps.

    a. Delete the previous polygon interior.

    b. Individually select each horizontal and vertical segment representing the height and the width of the step and choose **Construct** → Midpoint.

    c. Choose **Draw** → Line Segment and construct the additional steps.

    d. Select pairs of the new adjoined segments and choose **Constrain** → Perpendicular.

    e. Choose **Draw** → Polygon, and plot the new polygon for this staircase.

Q4. What is the length of the stairs if we replace each step with two steps of equal heights and widths?

A4: *There are total of 4 steps now, each step has width 0.5 ft and height 0.5 ft, so total length of the stairs is still 4 ft.*

4. Verify your answer using the software.

    a. Select the polygon and choose **Calculate** → Perimeter.

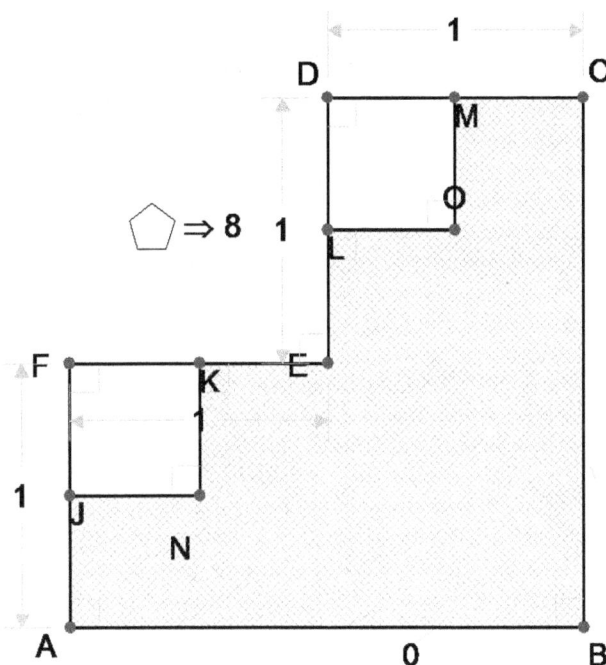

Q5. What is the height and width of a step if you have *2n* steps total? Fill in the table below to help you answer the question.

A5: Consider the pattern:

Number of pairs of steps	Height of one step
1	*1*
2	*1/2*
4	*1/4*
n	*1/n*

So, if we have n pairs of steps, we must have 2n steps, and the height and width of each step is 2/n.

Q6. What is the length of the staircase as the number of steps increases without bound?

A6: $L_n = 2n \cdot \dfrac{2}{n} = 4$, *so* $\lim\limits_{n \to \infty} L_n = \lim\limits_{n \to \infty} 4 = 4$.

Q7. Compare the stairs problem with a case of the inscribed regular n-gon as the number of sides gets infinitely many. In both problems one geometric figure approaches the other one sufficiently close. What happens to the perimeter of the inscribed regular n-gon as the number of sides increases infinitely?

Case of stairs	Case of a circle

A7: The perimeter of the regular n-gon converges to the circumference of the circle, but the length of the stairs does not converge to length of the hypotenuse of the triangle. These two examples illustrate that as one geometric figure approaches another geometric figure sufficiently close it does not guarantee that the corresponding measurements of these two figures also approach each other sufficiently close.

Students can illustrate the process of "smoothing" the stairs by setting up the staircase with a variable n to represent the number of steps. Then use a slider in the Variables menu to setup a large upper limit for n and observe what happens as n gets larger.

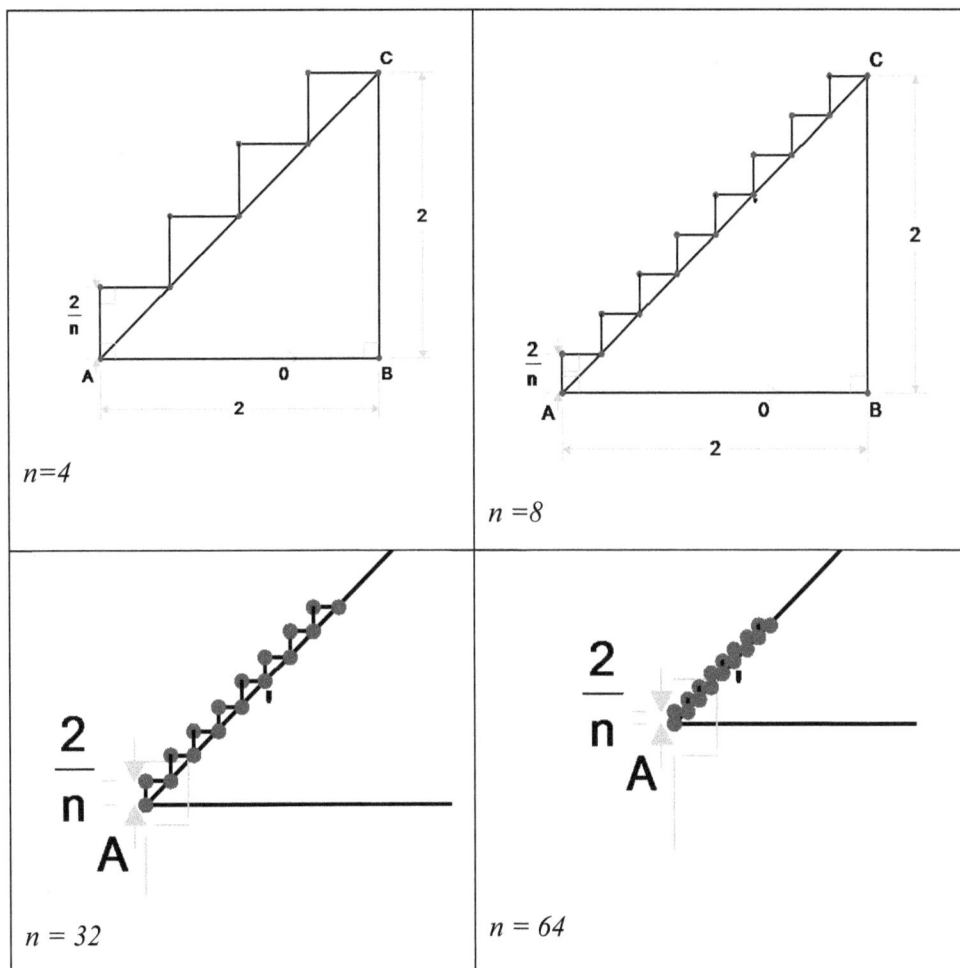

In order to setup this illustration, students can follow these steps:

1. Choose **Draw** → *Line Segment and draw two segments representing one step starting from the point A.*

2. Using **Constrain** → *Perpendicular to make the step perpendicular to the base of the staircase and the top of the step horizontal to the base.*

3. Use **Constrain** → *Distance to set the height of the step to be 2/n.*

---

4. *Use **Construct** → Translation and translate the step up the stairs. Then select two steps and translate them up the stairs.*

5. *Repeat that 2-3 more times, and you are ready to illustrate the process of "smoothing" the stairs.*

Infinite Stairs

Exploration 6.1: A staircase leads from point A to point C and has two steps, each one is 1 ft high and 1 ft wide. You decide to create a new staircase that will still lead from point A to point C, but with 4 steps so that each step is now 0.5 ft high and 0.5 ft wide. You doubled the number of steps while keeping their heights equal to their widths. You are interested in the total length of the stairs defined as the total sum of the heights and widths of all the steps. If you continue this process, what is the total length of the stairs as the number of steps increases without bound?

UP THE STAIRCASE

Q1. What do you think the length of the stairs will be when the number of steps increases without bound?

1. Draw an L-shaped staircase with two steps as shown on the diagram below.

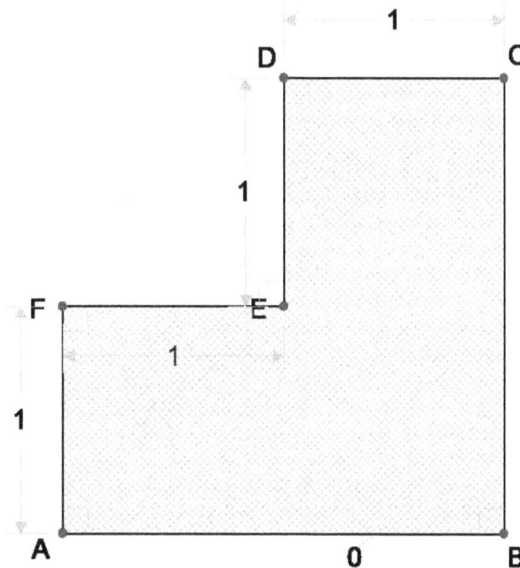

Q2. What is the total height and the total width of this set of stairs?

Q3. What is the total length of the stairs measured as the sum of heights and widths of the steps?

2. Verify your answer using the software.

3.   Now draw a staircase with 4 steps.

Q4. What is the length of the stairs if we replace each step with two steps of equal heights and widths?

4.   Verify your answer using the software.

Q5. What is the height and width of a step of the stairs if you have $2n$ steps total? Fill in the table below to help answer the question.

Number of pairs of steps	Height of one step
1	
2	
4	
n	

Q6. What is the length of the staircase as the number of steps increases without bound?

Q7. Compare the stairs problem with a case of the inscribed regular n-gon as the number of sides gets infinitely many. In both problems one geometric figure approaches the other one sufficiently close. What happens to the perimeter of the inscribed regular n-gon as the number of sides increases infinitely?

Case of stairs	Case of circle
Case of stairs	Case of circle

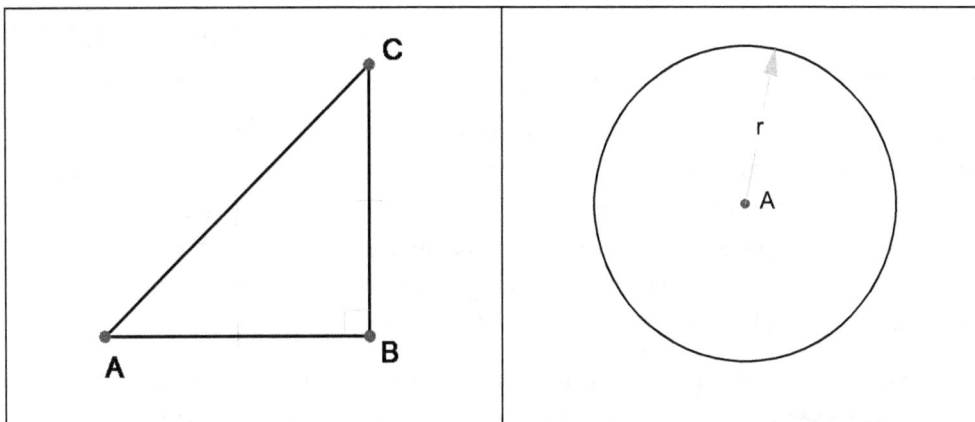

6.2 The Snowman Problem

Exploration 6.2: Given an equilateral triangle with side length a. A circle is inscribed in the triangle. Place a 2nd circle in the space between one of the vertices of the equilateral triangle and the 1st circle, so that it is tangent to the 1st circle and to two sides of the equilateral triangle. Repeat the process and place a 3rd circle in the space between the 2nd circle and the same vertex of the triangle, so that it is tangent to the 2nd circle and to the sides of the equilateral triangle. This process continues without limit.

1. What is the area of n circles that are constructed by the method described above? If n increases without bound, will the area of all the circles be bounded?

2. If we complete the same process at all vertices of the triangle, what part of the equilateral triangle is filled by the circles?

SUMMARY

Mathematics Objectives:

*   Derive recursive and closed form equations for the nth term of the geometric series

*   Find the partial sum of the geometric series and investigate its convergence.

Vocabulary:

*   Geometric sequence

*   Geometric series

*   Bounded series

*   Convergence

Pre-requisites:

*   Area of a circle

*   Sigma notation

*   Sum of geometric series

Problem Notes:

- The Snowman problem provides students with a visual and algebraic way to investigate a geometric series.

- Students first develop a recursive relationship for the area of the nth circle using the symbolic output of areas provided by the software. Thus, they establish the fact that the sequence of circle areas is geometric.

- Students then find the expression for the nth term of the series and setup the general expression for the series using sigma notation. They analyze the series for convergence using the formula for the sum of the geometric series.

Technology skills:

- Draw: polygon, circle

- Constrain: distance, tangent

- Calculate: area

Extension:

Given an equilateral triangle ABC, construct an equilateral triangle with vertices at midpoints of the sides of triangle ABC. Color its interior. Repeat this process again and construct one equilateral triangle in each of three empty triangles. Color interior of these three triangles. Continue this process indefinitely. (This process will create a fractal known as Sierpinski triangle).

1. Prove that the union of the colored triangles fills the triangle ABC.

2. How many steps do you need to have the area of all triangles to exceed 99% of the area of ABC?

STEP-BY-STEP INSTRUCTIONS

BUILDING A SNOWMAN

1. Draw an equilateral triangle with side a.

   a. Use **Toggle grid and axes** to hide the axes and the grid.

   b. Choose **Draw** → Polygon and plot a triangle ABC. Hide the triangle interior.

   c. Select segment AB and choose **Constrain** → Distance. Type a in the open edit box

    d. Repeat that for segments BC and AB.

2.   Draw a circle inscribed in the triangle.

    a. Choose **Draw** → Circle. Plot a circle inside the triangle.

    b. Select the circle and side AB and choose **Constrain** → Tangent.

    c. Repeat the process for each side of the triangle.

Q1. What is the area of the inscribed circle?

*A1: Students can use the software to answer this question by selecting the circle and choosing **Calculate** → Area. Or they can use geometry to derive the radius of the inscribed circle: $A_1 = \dfrac{\pi a^2}{12}$.*

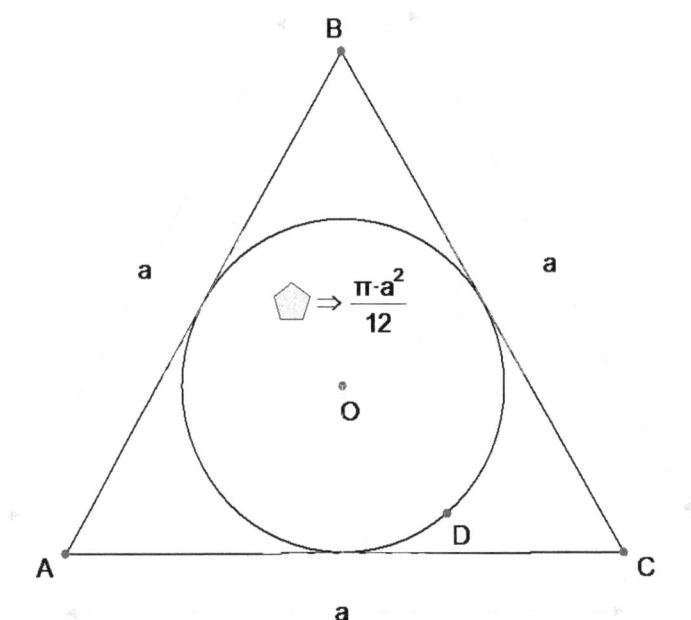

3.   Construct a 2nd circle between a vertex and the 1st circle, such that the circle is tangent to the 1st circle and to both sides of the vertex of the triangle.

    a. Choose **Draw** → Circle. Plot the 2nd circle in the region above the first circle near vertex B.

    b. Select both circles and choose **Constrain** → Tangent.

c. Select the 2nd circle and the side AB and choose **Constrain** → Tangent.

d. Similarly make side BC of the triangle tangent to the new circle.

Q2. What is the area of the 2nd circle? Use the software to find the answer.

A2: Students use the software to answer this question by selecting the circle and choosing
Calculate → *Area.* $A_2 = \dfrac{\pi a^2}{108}$.

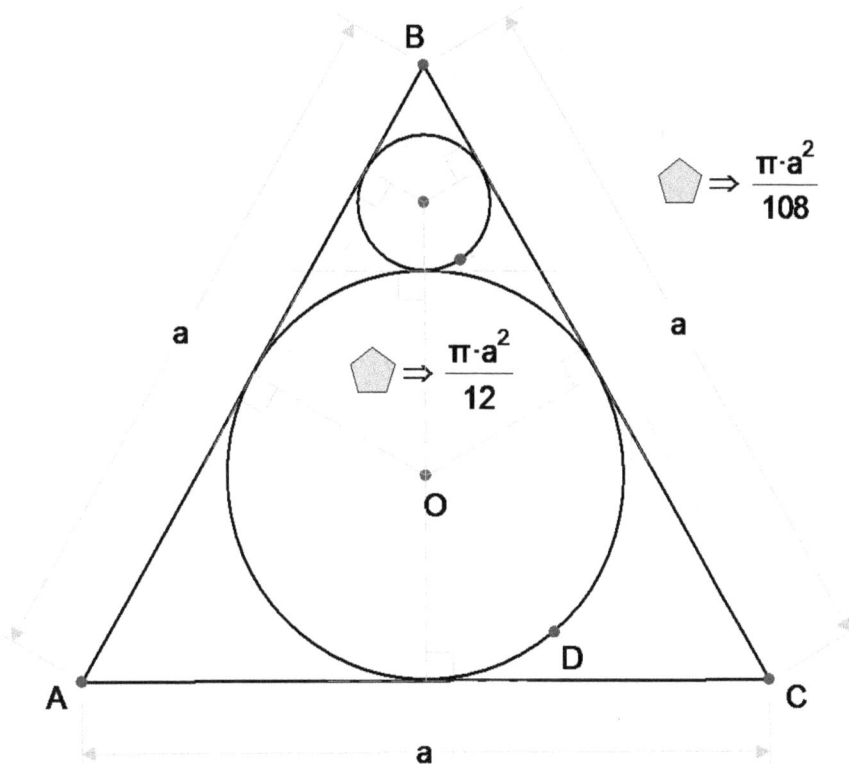

4. Similarly construct a 3rd circle between the 2nd circle and the same vertex B that is
    tangent to the 2nd circle and to both sides of the equilateral triangle. (If your picture
    does not look like a snowman, go back and check your construction.)

a. Choose **Draw** → Circle. Plot the 3rd circle in the region above the second circle
    near vertex B.

b. Select the 2nd and 3rd circles and choose **Constrain** → Tangent.

c. Select the 3rd circle and side AB and choose **Constrain** $\rightarrow$ Tangent.

d. Similarly make the side BC of the triangle tangent to the circle.

Q3. What is the area of the new 3rd circle? Use *Gx* to find the answer.

A3: Students use the software to answer this question by selecting the circle and choosing

Calculate $\rightarrow$ *Area.* $A_3 = \dfrac{\pi a^2}{972}$.

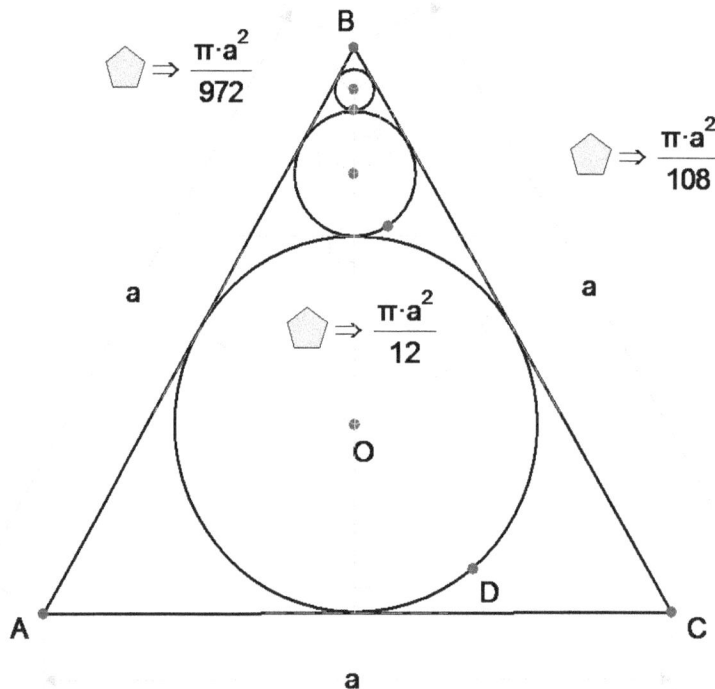

Q4. What is the ratio of the areas $\dfrac{A_2}{A_1}, \dfrac{A_3}{A_2}$?

A4: $\dfrac{A_2}{A_1} = \dfrac{\pi a^2}{108} \cdot \dfrac{12}{\pi a^2} = \dfrac{1}{9}$; $\dfrac{A_3}{A_2} = \dfrac{\pi a^2}{972} \cdot \dfrac{108}{\pi a^2} = \dfrac{1}{9}$. *The ratios are the same, so this is a geometric sequence.*

TO INFINITY AND BEYOND ... SNOWMEN

Q5. If we continue the process above, what is the area of the nth circle constructed?

A5: $A_n = \dfrac{1}{9}A_{n-1} = \dfrac{1}{9^2}A_{n-2} = \dfrac{1}{9^{n-1}}A_1$

Q6. Write the sum of the areas of n circles using Σ notation.

A6: $S_n = \displaystyle\sum_{k=1}^{n} \dfrac{\pi a^2}{12}\left(\dfrac{1}{9}\right)^{k-1} = \dfrac{3\pi a^2}{4}\sum_{k=1}^{n}\left(\dfrac{1}{9}\right)^k$

Q7. Analyze the convergence of the series when the number of circles increases without bound. If it exists, what is the geometric meaning of the sum of the series?

A7: *Since this is a geometric series,* $S = \displaystyle\lim_{n\to\infty} S_n = \dfrac{3\pi a^2}{4}\sum_{k=1}^{\infty}\left(\dfrac{1}{9}\right)^k = \dfrac{3\pi a^2}{4}\dfrac{\frac{1}{9}}{1-\frac{1}{9}} = \dfrac{3\pi a^2}{32}$. *This*

value is the limit of the sum of the areas of the circles at one vertex as the number of circles increases without bound.

Q8. If we repeat the same process at each vertex of the triangle, will the circles at the other vertices have the same areas as the original set of circles?

A8: *Students can construct the circles at the other two vertices and confirm that the areas are the same using the software.*

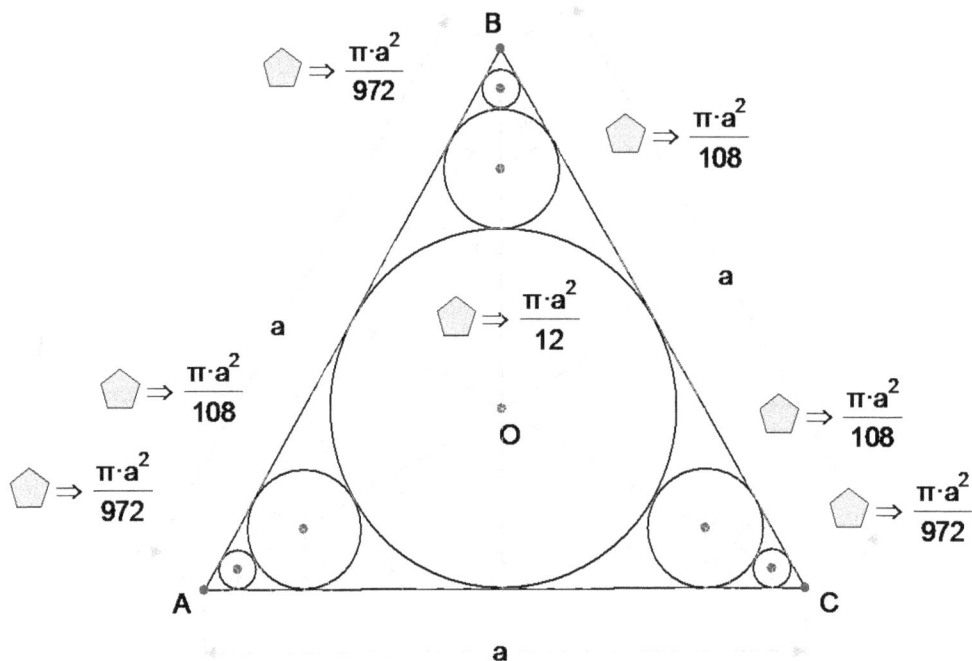

Q9. Setup a series for the area of all circles as their number increases without bound. Use Σ notation.

A9: $S = \dfrac{\pi a^2}{12} + \dfrac{\pi a^2}{36} \sum\limits_{n=0}^{\infty} \left(\dfrac{1}{9}\right)^n$, *or* $\dfrac{9\pi a^2}{4} \sum\limits_{n=1}^{\infty} \left(\dfrac{1}{9}\right)^n - \dfrac{\pi a^2}{6}$.

Q10. What part of the triangle is filled by the circles?

A10: $S = \dfrac{\pi a^2}{12} + \dfrac{\pi a^2}{36} \cdot \dfrac{1}{1 - \dfrac{1}{9}} = \dfrac{\pi a^2}{12} + \dfrac{\pi a^2}{36} \cdot \dfrac{9}{8} = \dfrac{11\pi a^2}{96}$,

$A = \dfrac{1}{2} a^2 \sin \dfrac{\pi}{3} = \dfrac{\sqrt{3} a^2}{4}$,

$\dfrac{S}{A} = \dfrac{11\pi}{96} \cdot \dfrac{4}{\sqrt{3}} \approx 0.83$.

The Snowman Problem

<u>Exploration 6.2:</u> Given an equilateral triangle with side a. A circle is inscribed in the triangle. Place a 2nd circle in the space between one of the vertices of the equilateral triangle and the 1st circle, so that it is tangent to the 1st circle and to the two sides of the equilateral triangle. Repeat the process and place a 3rd circle in the space between the 2nd circle and the same vertex of the triangle, so that it is tangent to the 2nd circle and to the sides of the equilateral triangle. This process continues without limit.

1. What is the area of n circles that are constructed by the method described above? If n increases without bound, will the area of all circles be bounded?

2. If we complete the same process at all vertices of the triangle, what part of the equilateral triangle is filled by the circles?

BUILDING A SNOWMAN

1. Draw an equilateral triangle with side a.

2. Draw a circle inscribed in the triangle.

Q1. What is the area of the inscribed circle?

3. Construct a 2nd circle between a vertex and the 1st circle, such that the circle is tangent to the 1st circle and to both sides of the vertex of the triangle.

Q2. What is the area of the 2nd circle? Use the software to find the answer.

4. Similarly construct a 3rd circle between the 2nd circle and the same vertex B that is tangent to the 2nd circle and to both sides of the equilateral triangle. (If your picture does not look like a snowman, go back and check your construction.)

Q3. What is the area of the new 3rd circle? Use Gx to find the answer.

Q4. What is the ratio of the areas $\dfrac{A_2}{A_1}, \dfrac{A_3}{A_2}$?

TO INFINITY AND BEYOND ... SNOWMEN

Q5. If we continue the process above, what is the area of the nth circle constructed?

Q6. Write the sum of the areas of n circles using Σ notation.

Q7. Analyze the convergence of the series when the number of circles increases without bound. If it exists, what is the geometric meaning of the sum of the series?

Q8. If we repeat the same process at each vertex of the triangle, will the circles at the other vertices have the same areas as the original set of circles?

Q9. Setup a series for the area of all circles as their number increases without bound. Use Σ notation.

Q10. What part of the triangle is filled by the circles?

6.3 Trigonometric Delight

Exploration 6.3: On the x-axis point P_0 is at the origin and point P_1 is at $x = 1$. Draw a line through P_0 that forms an acute angle α with positive direction of the x – axis. Draw a perpendicular to the x-axis from the point P_1. Let the intersection of the perpendicular and the line through P_0 be Q_0. Now draw a line through P_1 that forms an angle α with the positive direction of the x – axis. Plot point Q_1 on this line so that $P_1Q_1 = P_0P_1$. Drop a perpendicular from Q_1 to the x-axis and mark the point P_2.

Repeat the same process to find Q_2 and P_3: Draw a line through P_2 that forms an angle α with the positive direction of the x – axis. Plot point Q_2 on this line so that $P_2Q_2 = P_1P_2$. Drop a perpendicular from Q_2 to the x-axis and mark the point P_3.

1. If you continue this process in this manner, what can you say about the length P_0P_n as n increases without bound?

2. Analyze this problem and find its solution geometrically.

3. Analyze this problem and find its solution analytically.

SUMMARY

Mathematics Objectives:

- Find the partial sum of geometric series based on the properties of similar triangles.

- Use geometric and algebraic considerations to analyze convergence of the geometric series.

- Provide geometric interpretation for the sum of the series.

- Investigate the series for positive and negative values of geometric ratio.

Vocabulary:

- Geometric series

- Convergence

- Partial sum

Pre-requisites:

- Similar triangles

- Right triangle trigonometry relationships

- Sum of a geometric series.

Problem Notes:

- Every infinite geometric progression can be constructed geometrically, and its sum can be found graphically using only a straightedge and a compass. A geometric series with the ratio r converges if $-1 < r < 1$. On the other hand, any number between -1 and 1 can be represented by a cosine of an angle. Thus, for any geometric series we can have $r = \cos \alpha$ providing the basis for this investigation.

- Students first construct a sequence of similar right triangles with their bases on the x-axis and a given acute angle and justify that the vertices of these triangles lie on the same line that intersects the x-axis. They find the combined length of the bases of these triangles in terms of the given angle as the number of triangles increases without a bound.

- Then students setup the expression for the combined length of the bases for the n triangles in terms of the cosine of the given angle using right triangular trigonometry and similarity relationships. Using the limit procedure, students find the sum of an infinite series and determine that it is equal to the length of the segment found geometrically.

- What makes this problem so interesting is that the sum of an infinite series can be found geometrically using only a straightedge and a compass. We just construct a series of similar right triangles with vertices at P_0, P_1, P_2, P_3, …, along the x-axis and Q_0, Q_1, Q_2, …, along the line that intersects the x-axis at point P. Then the length of the segment P_0P provides the sum of infinite series, $\displaystyle\sum_{k=0}^{\infty} \left| \overline{P_k P_{k+1}} \right|$.

- Not only does this provide a geometric interpretation of a geometric series, it also allows students to see what happens to the sum of an infinite series if we change the common ratio of the series by varying the angle created by the two lines.

Technology skills:

- Draw: point, line, line segment

- Constrain: point proportional long the curve, direction of a line, congruent segments, perpendicular

- Calculate: distance

- Construct: parallel, intersection

Extension:

Draw points $O(0,0)$ and $A_1(1,0)$. In order to find point A_2 construct a segment $A_1 A_2 \perp OA_1$ with the length $\overline{OA_2} = a\overline{OA_1}$, where $a > 1$. Repeat this process the same way, so $A_n A_{n+1} \perp OA_n$ and $\overline{OA_{n+1}} = a\overline{OA_n}$. You notice that the point A_{12} happened to lie on the x-axis so that you completed the full circle.

   a. Find the x-coordinate of the point A_{12}.

   b. Find the total combined area of the triangles $OA_1 A_2$, $OA_2 A_3$, …, $OA_{11} A_{12}$.

   c. Find the parameter c such that all points A_1, A_2, …A_{12} lie on a spiral with the equation $r = c^\theta$.

STEP-BY-STEP INSTRUCTIONS

THE GEOMETRIC APPROACH

1.   Complete the 1st step of the construction to locate points P_1 and Q_0.

   a. Use **Toggle grid and axes** to show the axes without the grid.

   b. Choose **Draw** → Point and plot a point at the origin. Label it P[0].

   c. Choose **Draw** → Line and plot a line through the origin. Select the line and choose **Constrain** → Direction. Type α.

   d. Choose **Draw** → Point and plot a point on the x-axis. Label it P[1].

   e. Select P_1 and the x-axis, and choose **Constrain** → Proportional. Type 1 in the open edit box.

f.   Choose **Draw** → Segment and plot a segment from P_1 to the line. Select the segment and the x-axis and choose **Constrain** → Perpendicular. Label the point on the line Q[0].

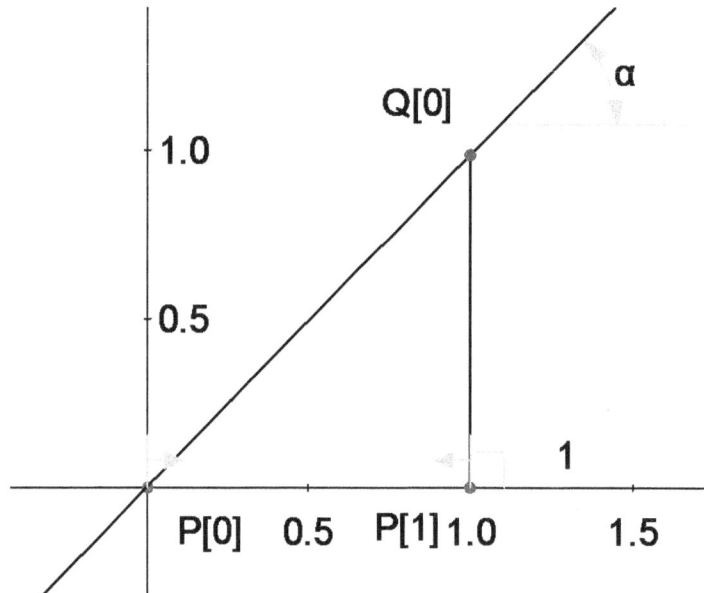

2.   Complete the 2nd step of the process and find points P_2 and Q_1

a.   Select point P_1 and the line P_0Q_0 and Choose **Construct** → Parallel

b.   Choose **Draw** → Segment and plot segment P_0P_1 and a new segment P_1Q_1 where Q_1 is a point on the parallel line constructed above. Hide the parallel line.

c.   Select both segments and choose **Constrain** → Congruent.

d.   Choose **Draw** → Segment and plot a segment from Q_1 to the x-axis. Select the segment and the x-axis and choose **Constrain** → Perpendicular. Label the point on the x-axis P[2].

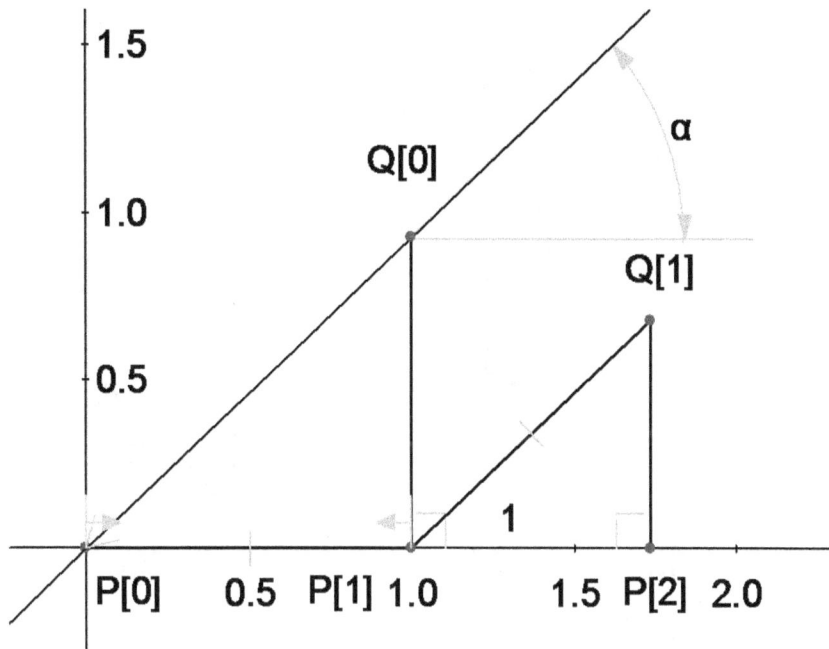

3. Complete one more step and locate points P_3 and Q_2

    a. Select the point P_2 and the line P_0Q_0 and Choose **Construct** → Parallel

    b. Choose **Draw** → Segment and plot the segment P_1P_2 and the segment P_2Q_2 along the parallel line constructed above. Hide this line.

    c. Select both segments and choose **Constrain** → Congruent.

    d. Choose **Draw** → Segment and plot a segment from Q_2 to the x-axis. Select the segment and the x-axis and choose **Constrain** → Perpendicular. Label the point on the x-axis P[3].

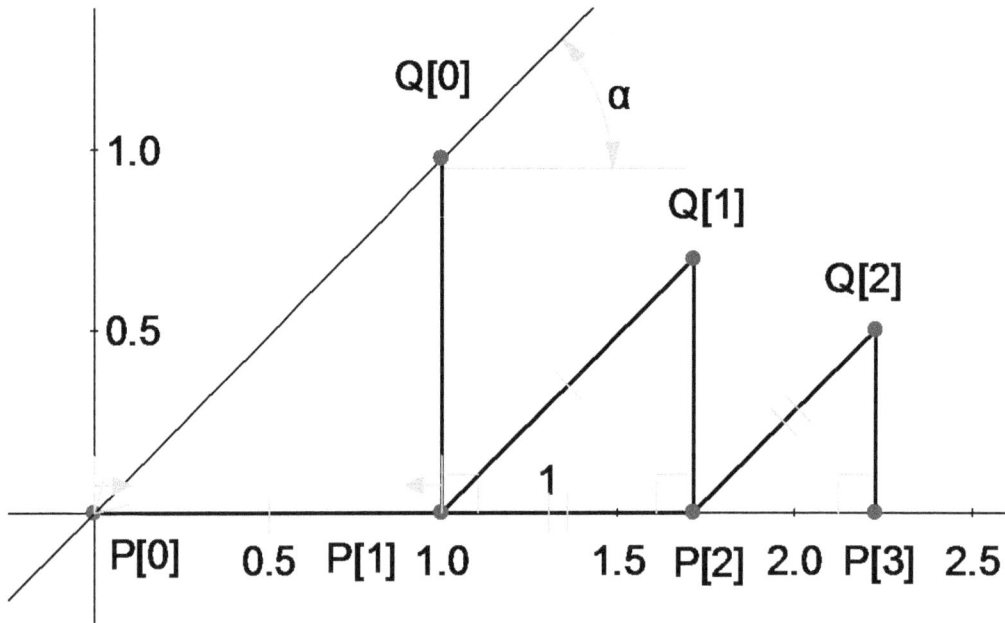

Q1. How are the triangles $\Delta P_0 Q_0 P_1$, $\Delta P_1 Q_1 P_2$, and $\Delta P_2 Q_2 P_3$ related?

A1: *These triangles are similar since all the corresponding angles are equal. The similarity*
ratio can be found as $\dfrac{P_1 Q_1}{P_0 Q_0} = \dfrac{1}{1/\cos\alpha} = \cos\alpha$.

4. Confirm your result with the help of the software.

   a. Choose **Draw** $\to$ Segment, and plot segments $P_0 Q_0$ and $P_2 P_3$.

   b. In order to calculate $P_0 Q_0$, $P_1 Q_1$, $P_2 Q_2$, select one segment at a time and choose **Calculate** $\to$ Distance.

   c. Similarly measure $P_0 P_1$, $P_1 P_2$, $P_2 P_3$.

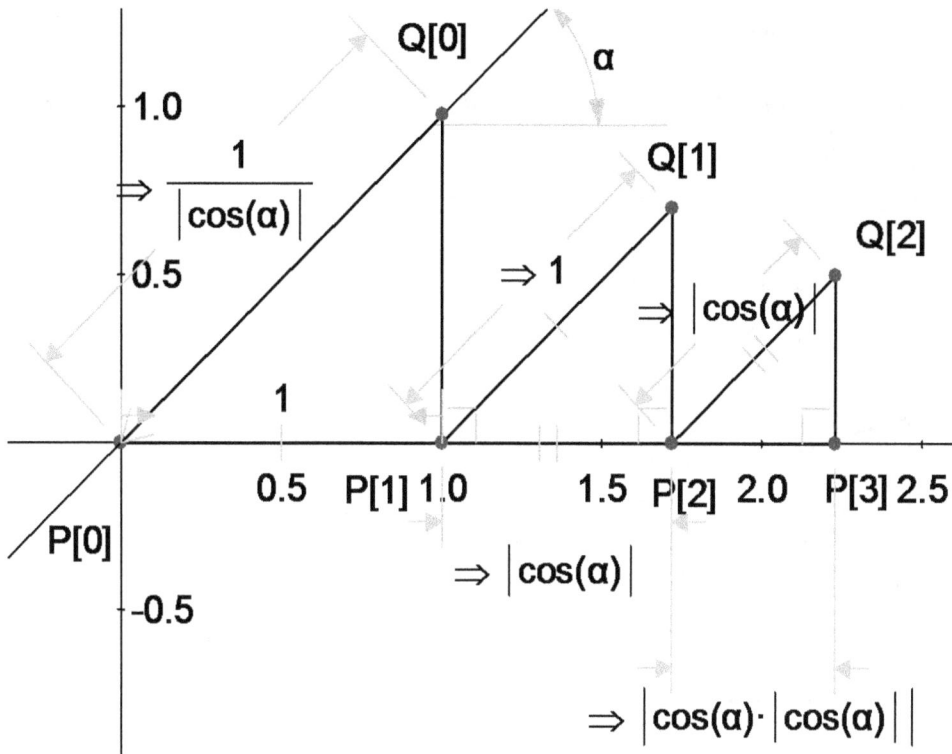

Q2. What can you say about the relative position of points Q_0, Q_1, Q_2?

A2: These points lie on a line. To justify that they are collinear, find the slope of the line through points Q_0 and Q_1: $m_1 = \dfrac{\sin\alpha - \tan\alpha}{\cos\alpha} = -\tan\alpha\dfrac{1-\cos\alpha}{\cos\alpha}$, and find slope of the line through points Q_1 and Q_2: $m_2 = \dfrac{\sin\alpha\cos a - \sin\alpha}{\cos^2\alpha} = -\tan\alpha\dfrac{1-\cos\alpha}{\cos\alpha}$. Since both lines have the same slope and pass through the same point, they are the same line.

5. Find the point of intersection between the line Q_0Q_1 and the x – axis and label this point P.

    a. Choose **Draw** → Line, and plot a line through points Q_0Q_1.

    b. Select the line and the x-axis and choose **Construct** → Intersection. Label this point P.

Q3. What is the distance P_0P?

A3: *Let $P_0P = S$. Since $\triangle P_0Q_0P \approx \triangle P_1Q_1P$, then $\dfrac{P_0P}{P_0Q_0} = \dfrac{P_1P}{P_1Q_1} \Rightarrow \dfrac{S}{1/\cos\alpha} = \dfrac{S-1}{1}$. Solving for*

S we get: $S = \dfrac{1}{1-\cos\alpha} = \dfrac{|\sin(\alpha)|}{|\sin(\alpha)-\sin(\alpha)\cos(\alpha)|}$

6. Verify your answer using the software:

    a. Select points P_0 and P and choose **Calculate** → Distance

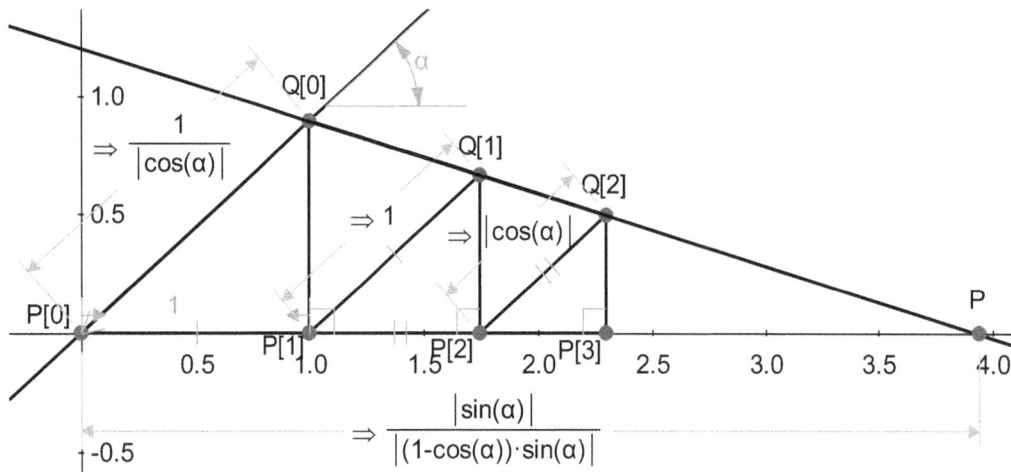

THE ANALYTIC APPROACH

Q4. Find distances P_0P_1, P_0P_2, and P_0P_3

A4: *$P_0P_1 = 1$; $P_0P_2 = 1 + \cos\alpha$, $P_0P_3 = 1 + \cos\alpha + \cos^2\alpha$.*

7. Calculate the distances P_0P_1, P_0P_2, P_0P_3 using the software.

    a. Select the endpoints of each interval and choose **Calculate** → Distance.

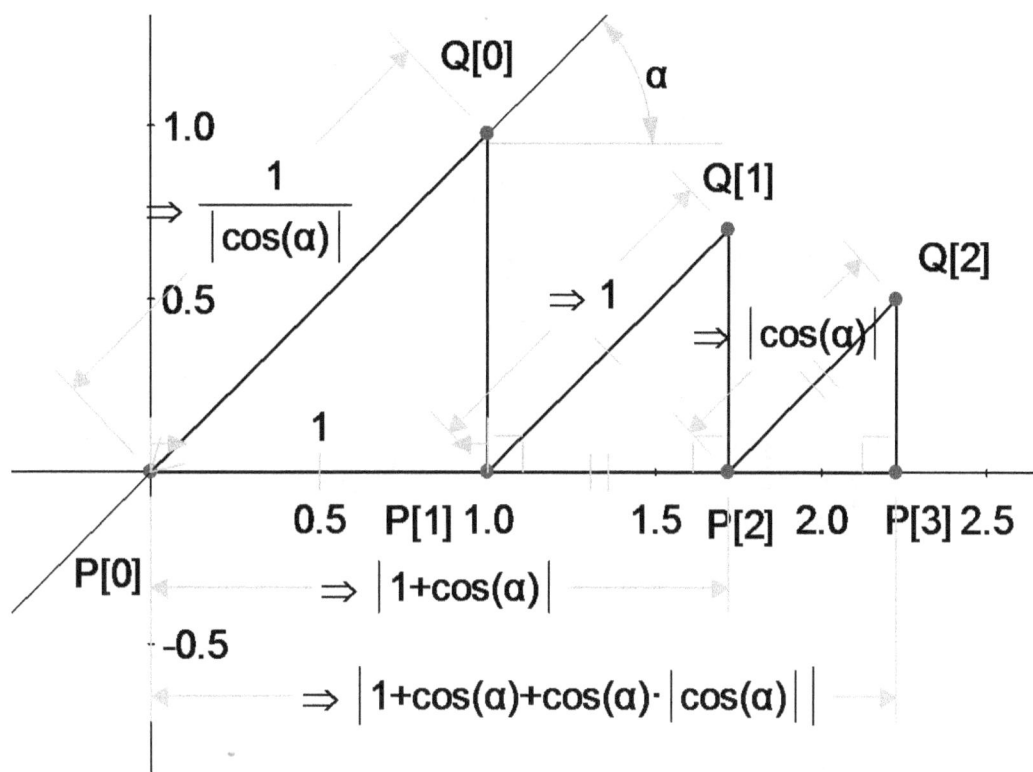

Q5. Based on your observations, find the distance P_0P_n.

A5: $P_0P_n = 1 + \cos \alpha + \cos^2 \alpha + \ldots + \cos^n \alpha.$

Q6. What happens to the point P_n and to the length of the segment P_0P_n as n increases without bound?

A6: Since $|\cos \alpha| < 1$, *the geometric series converges.* $L = \lim_{n \to \infty} P_0 P_n = \sum_{n=0}^{\infty} \cos^n \alpha = \dfrac{1}{1 - \cos \alpha}.$

Comparing the geometric and the analytic approach, we can see that $P_n \to P$ *as n increases without bound, since* $S = L$.

Q7. How does the sum of an infinite geometric series change when: a) the common ratio changes while remaining positive; b) the common ratio changes while remaining negative?

A7: Students can vary the value of the angle α *and observe the changes to the distance* P_0P *to provide an answer to this question.*

---

a. For $0 \le \alpha \le \dfrac{\pi}{2}$ the geometric ratio, $r = \cos \alpha > 0$. The larger the angle, the smaller r is, so the sum is smaller.

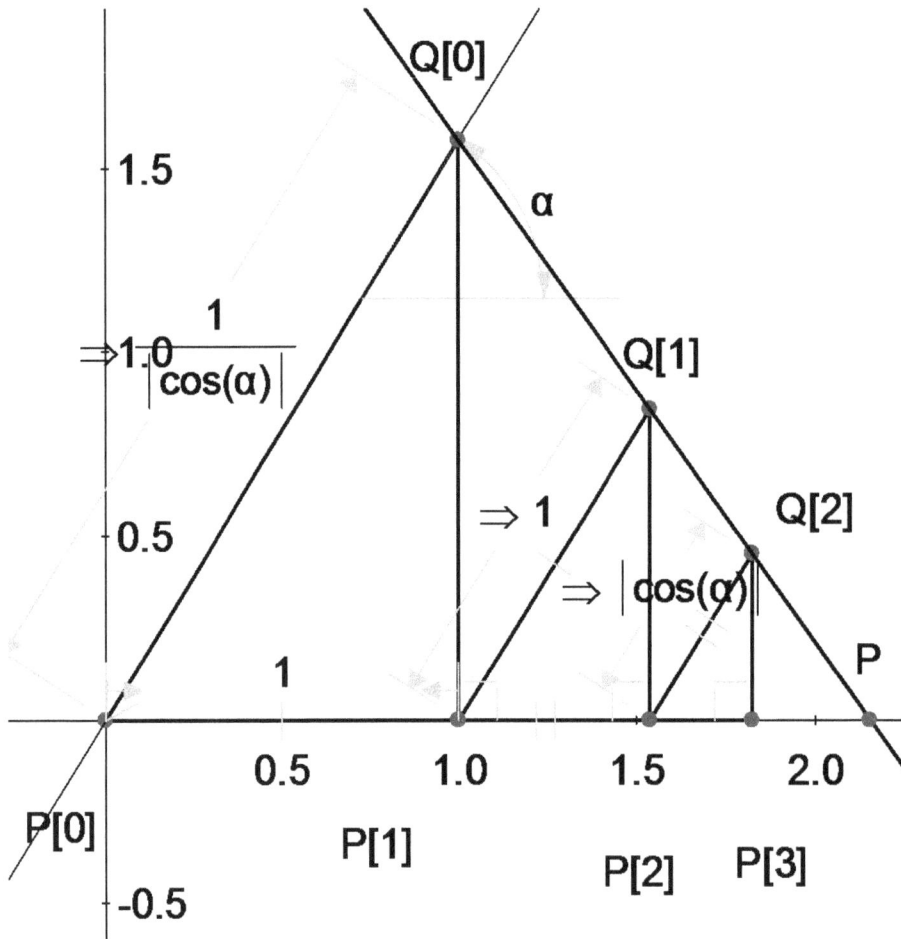

For $0 \le \alpha \le \dfrac{\pi}{2}$ the smaller the angle, the larger the geometric ratio is, so the sum is larger.

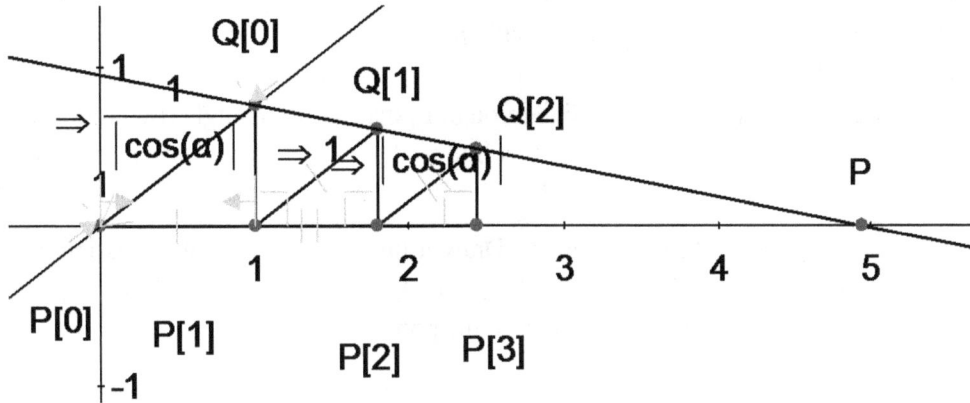

For $\dfrac{\pi}{2} \le \alpha \le \pi$, $r < 0$. In this case the similar triangles are nested inside each other so the geometric sequence is alternating. The sum of the series is less than 1.

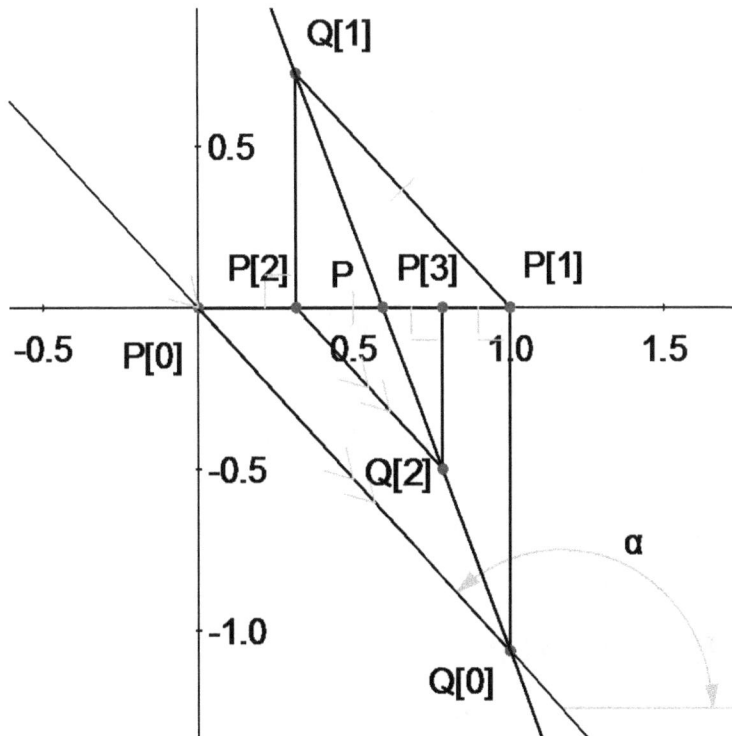

Trigonometric Delight

<u>Exploration 6.3:</u> On the x-axis point P_0 is at the origin and point P_1 is at $x = 1$. Draw a line through P_0 that forms an acute angle α with positive direction of the x – axis. Draw a perpendicular to the x-axis from the point P_1. Let the intersection of the perpendicular and the line through P_0 be Q_0. Now draw a line through P_1 that forms an angle α with the positive direction of the x – axis. Plot point Q_1 on this line so that $P_1Q_1 = P_0P_1$. Drop a perpendicular from Q_1 to the x-axis and mark the point P_2.

Repeat the same process to find Q_2 and P_3: Draw a line through P_2 that forms the angle α with the positive direction of the x – axis. Plot point Q_2 on this line so that $P_2Q_2 = P_1P_2$. Drop a perpendicular from Q_2 to the x-axis and mark the point P_3.

1. If you continue this process in this manner, what can you say about the length P_0P_n as n increases without bound?

2. Analyze this problem and find its solution geometrically.

3. Analyze this problem and find its solution analytically.

THE GEOMETRIC APPROACH

1. Complete the 1st step of construction to locate points P_1 and Q_0.

2. Complete the 2nd step of the process to find points P_2 and Q_1

3. Complete one more step and locate the points P_3 and Q_2

Q1. How are the triangles $\Delta P_0Q_0P_1$, $\Delta P_1Q_1P_2$, and $\Delta P_2Q_2P_3$ related?

4. Confirm your result with the help of the software.

Q2. What can you say about the relative position of points Q_0, Q_1, Q_2?

5. Find the point of intersection between the line Q_0Q_1 and the x – axis and label this point P.

Q3. What is distance P_0P?

6. Verify your answer using the software:

THE ANALYTIC APPROACH

Q4. Find distances P_0P_1, P_0P_2, and P_0P_3.

7.   Calculate the distances P_0P_1, P_0P_2, P_0P_3 using the software.

Q5. Based on your observations, find the distance P_0P_n.

Q6. What happens to the point P_n and to the length of the segment P_0P_n as n increases without bound?

Q7. How does the sum of an infinite geometric series change when a) the common ratio changes while remaining positive, and b) the common ratio changes while remaining negative.

6.4 Converging or Diverging?

<u>Exploration 6.4:</u> Given a right triangle with legs 1 and 1/2. Construct a 2nd right triangle that shares the leg of length 1/2 and has another leg 1/3 that is not on the same line as leg 1. The triangles don't overlap. Construct a 3rd triangle that shares the leg of length 1/3 of the 2nd triangle, second leg 1/4, that doesn't lie on the same line as ½ leg, and the triangles don't overlap. Continue this process *n* times.

1. What is the area of the polygon that is formed by the *n* triangles? If *n* increases without bound, will the area of the figure formed by the triangles be bounded?

2. What is the length of the polygonal chain formed by the legs of the triangles? If *n* increases without bound, will the length of the polygonal chain be bounded?

SUMMARY

<u>Mathematics Objective:</u>

- To discover the behavior of telescoping and harmonic series.

<u>Vocabulary:</u>

- Convergence
- Divergence
- Harmonic Series
- Telescopic Series
- Polygonal chain

<u>Pre-requisites:</u>

- Improper integral
- Sigma notation
- Partial sum of series
- Integral test of series convergence

Problem Notes:

- First students analyze the area of a polygon formed by the adjoined right triangles. With the help of the software students develop a recursive relationship for the nth term of the series, so they may discover the series and determine its convergence. This area is a visual representation of the partial sum of the series $\dfrac{1}{2}\sum_{k=1}^{n}\dfrac{1}{n(n+1)}$, known as the telescopic series.

- In the second part of the exploration, students analyze the length of the polygonal chain formed by the legs of the right triangles. This length is a partial sum of the series $\sum_{k=1}^{n}\dfrac{1}{n}$, known as the harmonic series. Since the polygonal chain is completely within the figure with converging area, students may intuitively assume that the length of the polygonal chain is also converging; however, when they attempt to prove convergence, they will discover that the harmonic series is, in fact, diverging.

Technology skills:

- Draw: polygon, function, arc, line segment
- Construct: polygon
- Constrain: point proportional along the curve, distance, perpendicular
- Calculate: area

Extension:

Find the area of the Koch snowflake.

STEP-BY-STEP INSTRUCTIONS

AREA OF A POLYGON FORMED BY N TRIANGLES

1. Draw a right triangle with legs 1 and 1/2.

    a. Use **Toggle grid and axes** to hide the axes and the grid.

    b. Choose **Draw** → Polygon and plot a triangle ABC. Select segments AB and BC and choose **Constrain** → Perpendicular.

---

    c.   Select segment AB and choose **Constrain** → Distance, type 1. Select segment BC and choose **Constrain** → Distance, type 1/2.

Q1.   What is the area of this triangle?

A1: $A_1 = \dfrac{1}{2}\left(1 \cdot \dfrac{1}{2}\right) = \dfrac{1}{4}$

2.   Verify your calculation with the software.

    a.   Select the triangle interior and choose **Calculate** → Area.

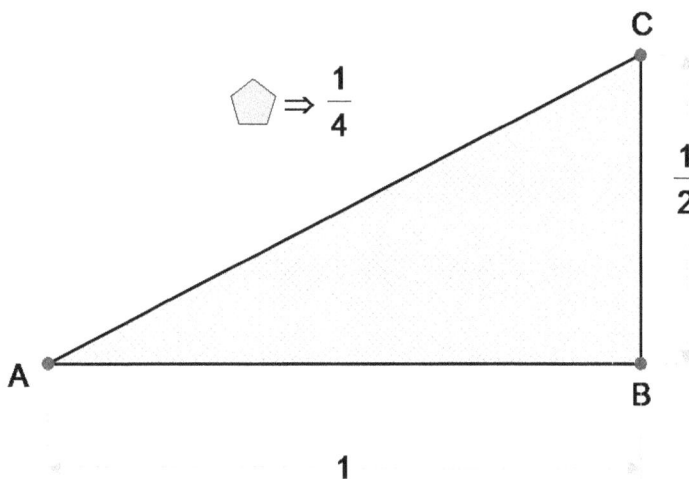

3.   Draw a 2nd right triangle using side BC with leg 1/2 and a new leg CD with side 1/3.

    a.   Choose **Draw** → Polygon and plot a triangle BCD. Select segments BC and CD, and choose **Constrain** → Perpendicular.

    b.   Select segment CD and choose **Constrain** → Distance, type 1/3.

Q2.   What is the area of the quadrilateral formed by ΔABC and ΔBCD?

A2: $S_2 = A_1 + A_2 = \dfrac{1}{2}\left(1 \cdot \dfrac{1}{2}\right) + \dfrac{1}{2}\left(\dfrac{1}{2} \cdot \dfrac{1}{3}\right) = \dfrac{1}{3}$

4.   Verify your calculation with the software.

    a.   Choose **Draw** → Polygon and plot the quadrilateral ABDC.

b. Select the quadrilateral interior and choose **Calculate** → Area.

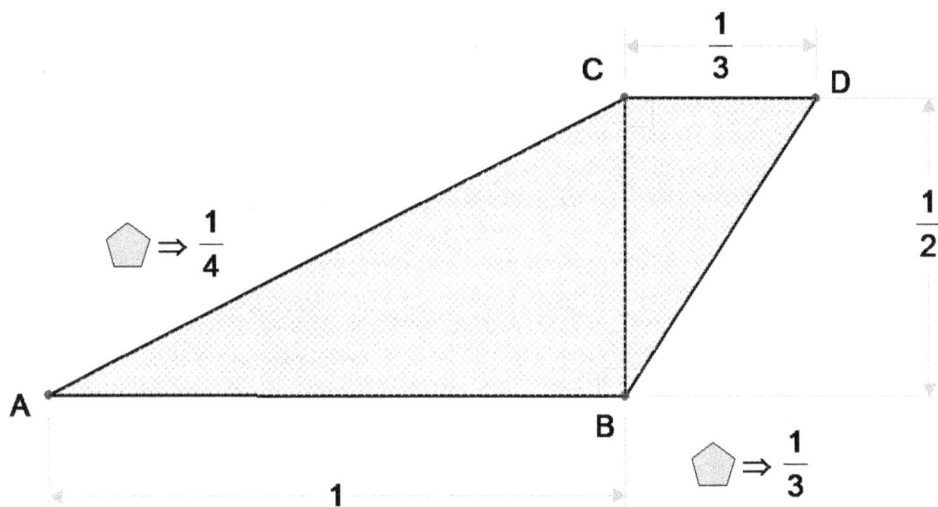

5. Continue the process by drawing a 3rd right triangle with leg CD (1/3) and a new leg DE with length 1/4.

a. Choose **Draw** → Polygon and plot a triangle CDE. Select segments CD and DE, and choose **Constrain** → Perpendicular.

b. Select segment DE and choose **Constrain** → Distance, type 1/4.

Q3. What is the area of the polygon formed by ΔABC, ΔBCD, and ΔCDE?

A3: $S_3 = A_1 + A_2 + A_3 = \dfrac{1}{2}\left(1 \cdot \dfrac{1}{2}\right) + \dfrac{1}{2}\left(\dfrac{1}{2} \cdot \dfrac{1}{3}\right) + \dfrac{1}{2}\left(\dfrac{1}{3} \cdot \dfrac{1}{4}\right) = \dfrac{1}{3} + \dfrac{1}{24} = \dfrac{3}{8}$

6. Verify your calculation with the software.

a. Choose **Draw** → Polygon and plot pentagon ABDEC.

b. Select pentagon interior and choose **Calculate** → Area.

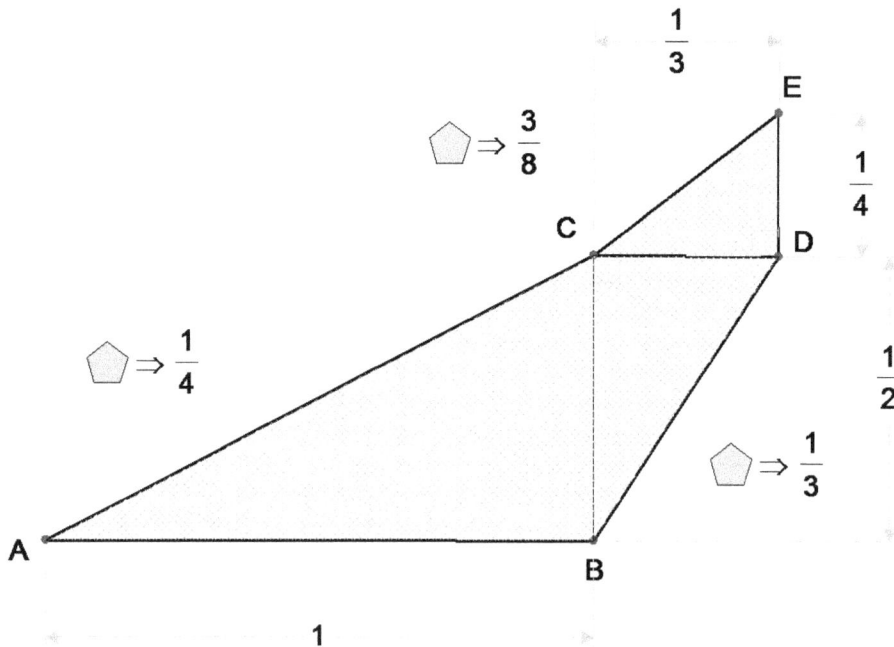

Q4. If you continue this process to the nth level, what are the measures of the legs of the nth triangle and what is its area?

A4: The legs of nth right triangle are measured $\dfrac{1}{n}$ and $\dfrac{1}{n+1}$. The area is $A_n = \dfrac{1}{2}\left(\dfrac{1}{n} \cdot \dfrac{1}{n+1}\right)$

Q5. What is the area of the polygon with n triangles?

A5: $S_n = A_1 + A_2 + A_3 + ... + A_n = S_3 + ... + A_n = \dfrac{3}{8} + ... + \dfrac{1}{2n(n+1)}$

7. Draw the nth triangle and verify your answers to questions 4 and 5 using the software.

a. Choose **Draw** $\rightarrow$ Polygon and plot a right triangle XYZ. Select segments XY and YZ, and choose **Constrain** $\rightarrow$ Perpendicular.

b. Select the segment XY and choose **Constrain** $\rightarrow$ Distance, type 1/n. Select the segment YZ and choose **Constrain** $\rightarrow$ Distance, type 1/(n+1)

c. Select the interior of this n^{th} triangle and choose **Calculate** $\rightarrow$ Area.

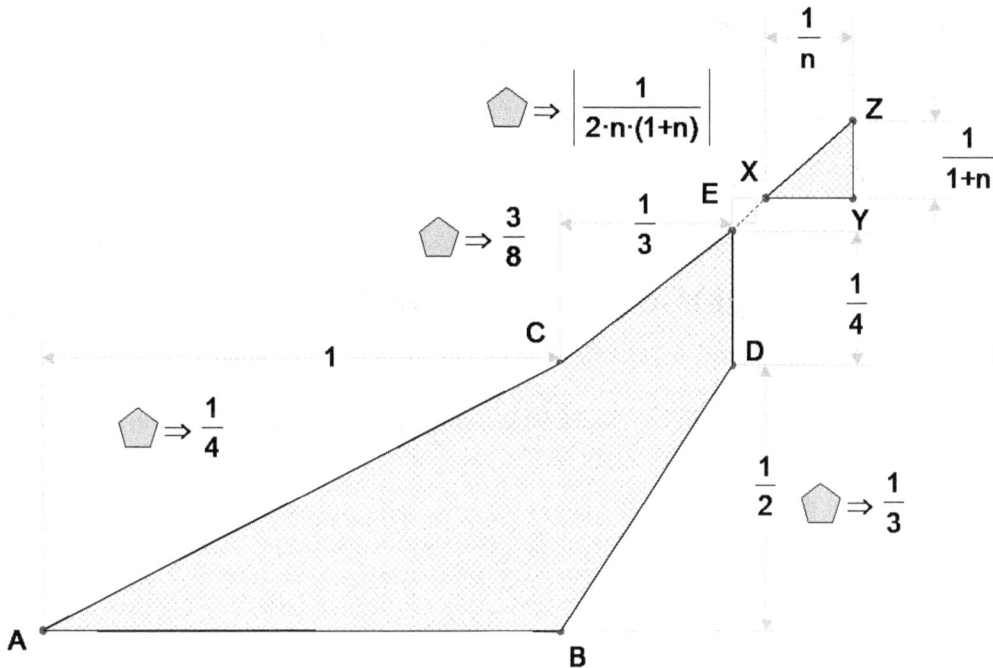

Q6. Observe the sequence of areas formed by the adjoined triangles. Does this sequence converge or diverge?

A6: *Since the first three terms they found are* $\dfrac{1}{4}, \dfrac{1}{3}, \dfrac{3}{8}, \dots$ *we know this is an increasing sequence; however, these three values are not sufficient to determine if the sequence is converging.*

Q7. Write the partial sum of the series representing the sum of the areas of the n triangles in Σ notation.

A7: $S_n = \dfrac{1}{2} \displaystyle\sum_{k=1}^{n} \dfrac{1}{k(k+1)}$

Q8. Determine if this series is converging or diverging.

A8: *This is a telescopic series,* $\dfrac{1}{n(n+1)} = \dfrac{1}{n} - \dfrac{1}{n+1}$, *so* $S_n = \sum\limits_{k=1}^{n} \dfrac{1}{2k(k+1)} = \dfrac{1}{2}\sum\limits_{k=1}^{n}\left(\dfrac{1}{k} - \dfrac{1}{k+1}\right)$,

then $s_n = \dfrac{1}{2}\left(1 - \dfrac{1}{2} + \dfrac{1}{2} - \dfrac{1}{3} + \dfrac{1}{3} - \ldots + \dfrac{1}{n} - \dfrac{1}{n+1}\right) = \dfrac{1}{2}\left(1 - \dfrac{1}{n+1}\right)$. *Thus the*

$\lim\limits_{n\to\infty} S_n = \dfrac{1}{2}\lim\limits_{n\to\infty}\left(1 - \dfrac{1}{n+1}\right) = \dfrac{1}{2}$, *so the area converges.*

LENGTH OF THE POLYGONAL CHAIN FORMED BY THE LEGS OF THE RIGHT TRIANGLES

Q9. Consider the polygonal chain ABCDE… formed by the legs of the right triangles. Write the partial sum of the series representing the sum of the lengths of n legs for $n = 1, 2, 3$, and a general n^{th} term.

A9: $L_1 = 1, L_2 = 1 + \dfrac{1}{2} = \dfrac{3}{2}, L_3 = 1 + \dfrac{1}{2} + \dfrac{1}{3} = \dfrac{11}{6}, L_n = 1 + \dfrac{1}{2} + \dfrac{1}{3} + \ldots + \dfrac{1}{n}.$

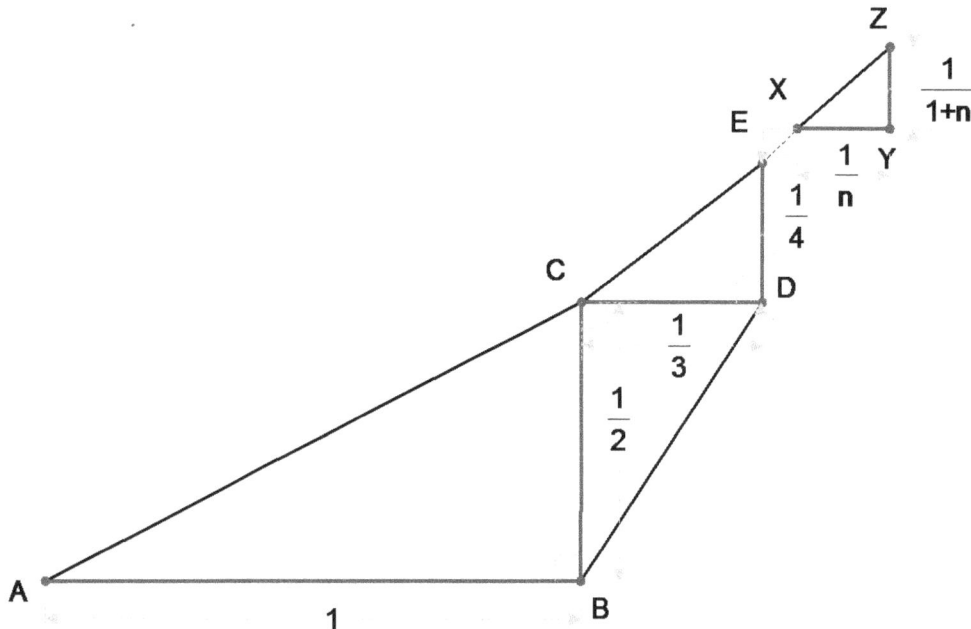

Q10. This polygonal chain is contained by the figure formed by the right triangles. What do you think is its length when $n \to \infty$?

A10: Answers will vary. Intuitively, students may perceive the length of the line to be converging, since it lies in the figure with converging area.

Q11. Write the partial sum of the series representing the sum of the lengths of the triangle legs in Σ notation.

A11: $L_n = \sum\limits_{k=1}^{n} \dfrac{1}{k}$

Q12. Determine if this series is converging or diverging.

A12: This is a harmonic series. Students can use the integral test and evaluate an improper

$$integral. \ \lim_{a \to \infty} \int_{1}^{a} \frac{dx}{x} = \lim_{a \to \infty} \ln x \Big|_{1}^{a} = \lim_{a \to \infty} \ln a - \ln 1 = \lim_{a \to \infty} \ln a = \infty. \ \textit{The limit does not exist,}$$

so the harmonic series diverges. Students can also verify their integration with the help of the software.

   a.   Choose **Draw** → Function. Select Cartesian, Y= 1/x, start 0 end 10.

   b.   Choose **Draw** → Points. Plot 2 points on the *x*-axis and 2 points on the graph of the function.

   c.   Select the 1st point on the axis and the x – axis, and choose **Constrain** → Point proportional. Type 1 in the edit box. Repeat this for the 2nd point you plotted on the axis, and type a.

   d.   Select the 1st point on the graph of the function and the curve, and choose **Constrain** → Point proportional. Type 1 in the edit box. Repeat this for the 2nd point on the curve, and type a.

   e.   Choose **Draw** → Arc and plot an arc from the 1st point to the 2nd point on the graph of the function.

   f.   Choose **Draw** → Segment and connect all the points to construct a curvilinear trapezoid with the arc on the graph of the function.

   g.   Select the arc and the segments and choose **Construct** → Polygon. Click on the interior of the polygon and choose **Calculate** → Area.

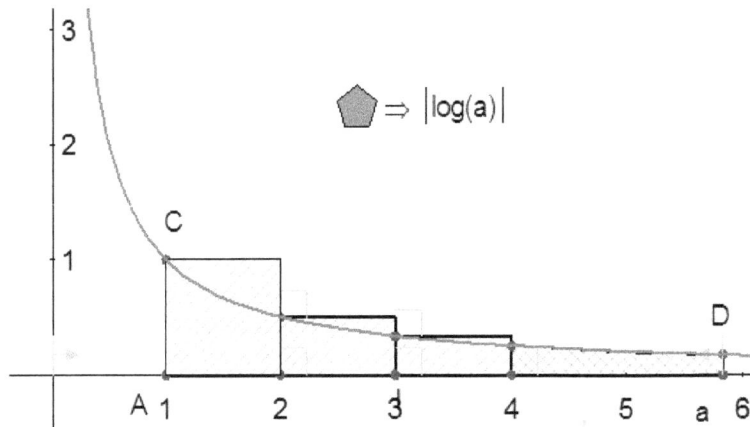

Note: the diagram above also shows the first three terms of the harmonic series as rectangles that are above the curve. Thus, $\sum_{n=1}^{\infty}\frac{1}{n} > \int_{1}^{\infty}\frac{dx}{x}$, which diverges.

Converging or Diverging?

<u>Exploration 6.4</u>: Given a right triangle with legs 1 and 1/2. Construct a 2nd right triangle that shares the leg of length 1/2 and has another leg 1/3 that is not on the same line as leg 1. The triangles don't overlap. Construct a 3rd triangle that shares the leg of length 1/3 of the 2nd triangle, second leg 1/4, that doesn't lie on the same line as ½ leg, and the triangles don't overlap. Continue this process n times.

1. What is the area of the polygon that is formed by the n triangles? If n increases without bound, will the area of the figure formed by the triangles be bounded?

2. What is the length of the broken line formed by the legs of the triangles? If n increases without bound, will the length of the broken line be bounded?

AREA OF A POLYGON FORMED BY N TRIANGLES

1. Draw a right triangle with legs 1 and 1/2.

Q1. What is the area of this triangle?

2. Verify your calculation with the software.

3. Draw a 2nd right triangle using side BC with leg 1/2 and a new leg CD with side 1/3.

Q2. What is the area of the quadrilateral formed by ΔABC and ΔBCD?

4. Verify your calculation with the software.

5. Continue the process by drawing a 3rd right triangle with leg CD (1/3) and a new leg DE with length 1/4.

Q3. What is the area of the polygon formed by ΔABC, ΔBCD, and ΔCDE?

6. Verify your calculation with the software.

Q4. If you continue this process to the nth level, what are the measures of the legs of the nth triangle and what is its area?

Q5. What is the area of the polygon generated by the n triangles?

7. Draw the nth triangle and verify your answers to questions 4 and 5 using the software.

Q6. Observe the sequence of areas of the polygons formed by adjoined triangles. Does this sequence converge?

Q7. Write the partial sum of the series representing the sum of the areas of the n triangles in Σ notation.

Q8. Determine if this series is converging or diverging.

LENGTH OF THE POLYGONAL CHAIN FORMED BY THE LEGS OF THE TRIANGLES

Definition: A **polygonal chain** (also known as **piecewise linear curve**) is a connected series of line segments.

Q9. Consider the polygonal chain ABCDE… formed by the legs of the right triangles. Write the partial sum of the series representing the sum of the lengths of n legs for $n = 1, 2, 3$, and the general n^{th} term.

Q10. This polygonal chain is contained by the figure formed by the right triangles. What do you think is its length when $n \to \infty$?

Q11. Write the partial sum of the series representing the sum of the lengths of the triangle legs in Σ notation.

Q12. Determine if this series is converging or diverging.

7. Parametric Equations and Polar Coordinates

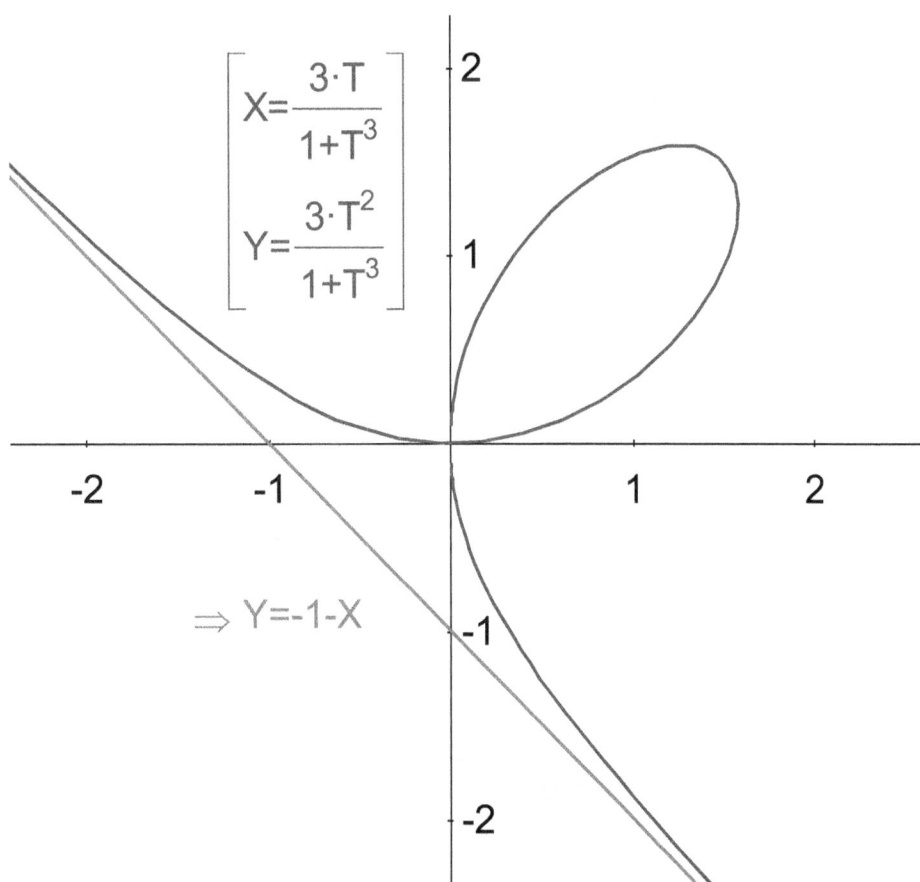

$$\begin{bmatrix} X = \dfrac{3 \cdot T}{1+T^3} \\[2ex] Y = \dfrac{3 \cdot T^2}{1+T^3} \end{bmatrix}$$

$\Rightarrow Y = -1 - X$

7.1 Folium of Descartes Using Parametric Equations

Exploration 7.1: In geometry the Folium of Descartes is a plane curve described by the equation $x^3 + y^3 - 3axy = 0$. The name of the curve comes from the Latin *folium* which means leaf, and as you will see, the curve forms a loop like a leaf. Using parametric equations, you will investigate the folium of Descartes with $a = 1$.

SUMMARY

Mathematics Objectives:

- Explore the symmetry of the Folium of Descartes using parametric equations.

- Explore the asymptote of the Folium of Descartes using parametric equations.

Vocabulary:

- Double Point

- Asymptote

Pre-requisites:

- Parametric equations

- Asymptotic behavior

- Limit at a point

- Limit at infinity

 Problem Notes:

- The Folium of Descartes, described by $x^3 + y^3 - 3xy = 0$, is a fascinating curve that is very difficult to graph in the Cartesian plane. This curve has a double point in the origin, symmetry about the line $y = x$, and a slant asymptote $y = -1 - x$. The curve passes through the origin twice at $\theta = 0$ and $\theta = \pi/2$ creating a double point.

- Students first derive parametric equations using the given Cartesian equation of the curve, substituting $y = xt$ in order to graph the curve. Since parametric equations are not used frequently in precalculus, it makes sense to discuss with the students how the parameter t is introduced. The meaning of the parameter $t = \dfrac{y}{x}$ is the tangent of the angle of the radius (or position) vector to a point on a curve.

- Students explore the symmetry of the curve both, algebraically and geometrically. They then explore the asymptotic behavior of the curve by analyzing the limits of $x(t)$ and $y(t)$ as $x \to \pm\infty$.

Technology skills:

- Draw: function, point

- Constrain: coordinates of a point, slope, point proportional along the curve

- Calculate: distance

- Construct: reflection.

Extension

Explore symmetry and asymptotic behavior of the Folium of Descartes for an arbitrary $a < 0$ using parametric equations.

STEP-BY-STEP INSTRUCTIONS

EXPLORING SYMMETRY

Q1. Can you graph the Folium of Descartes using its Cartesian equation: $x^3 + y^3 - 3xy = 0$? Explain.

A1: *Since it requires solving a cubic equation and graphing three different solutions, this curve is very difficult to graph using Cartesian equation. It is much easier to graph this curve using parametric equations or polar coordinates.*

Q2. Find parametric equations for this curve. (Hint: use $y = xt$, where t is a parameter)

A2: *Substitute $y = xt$ into the equation. $x^3 + (xt)^3 - 3x(xt) = 0 \Leftrightarrow x^2(x + xt^3 - 3t) = 0$, so solving for x and y we get: $x = \dfrac{3t}{1+t^3}$, $y = \dfrac{3t^2}{1+t^3}$.*

Note: It makes sense to discuss with the students how the parameter t is introduced. The meaning of the parameter $t = \dfrac{y}{x}$ is the tangent of an angle of the radius vector to a point on *the curve with the x-axis. The radius vector of a point is defined as a vector from the origin of the system of coordinates to the point. The angle of the radius vector is defined as the angle between the vector and the positive direction of the x-axis.*

1. Draw the curve using parametric equations.

    a. Use **Toggle grid and axes** to show the axes without the grid.

    b. Choose **Draw** → Function. For the Type select Parametric. In the X = and Y = prompts type the expressions for $x(t)$ and $y(t)$ you found. Select -30 for Start and 30 for End and press OK.

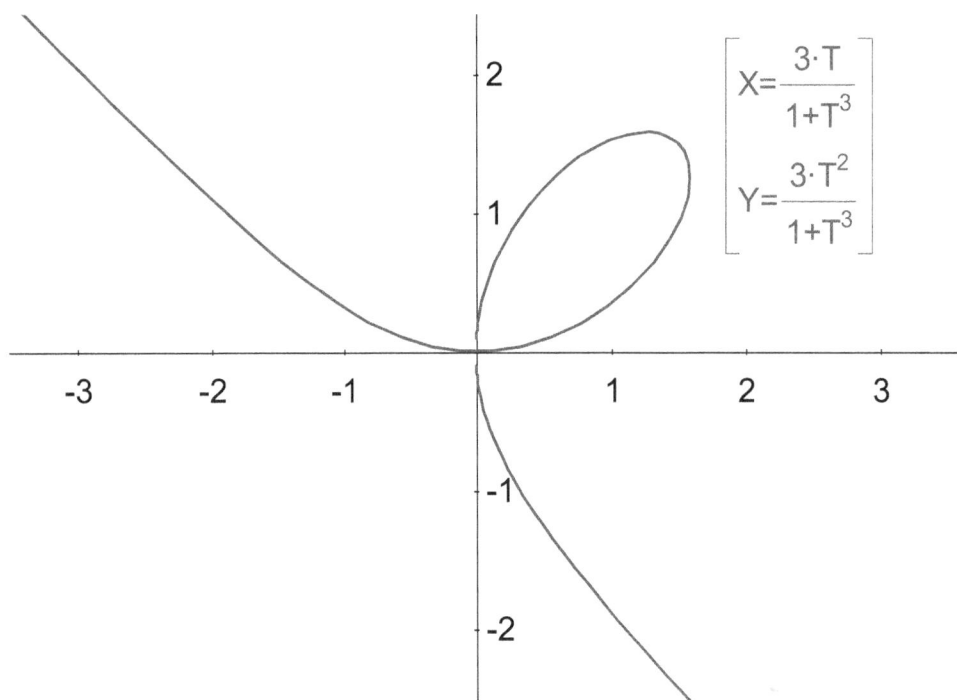

$$\left[\begin{array}{l} X = \dfrac{3 \cdot T}{1 + T^3} \\[2ex] Y = \dfrac{3 \cdot T^2}{1 + T^3} \end{array} \right.$$

Q3. What values of parameter t correspond to the different parts of the curve?

A3:   t < -1 corresponds to the wing in quadrant IV, -1 < t < 0 corresponds to the wing in quadrant II, and t > 0 corresponds to the loop in quadrant I. At t = 0 the curve is at the origin, and at t = -1, x and y are not defined.

*Note: students can use the software to plot a point on the curve and use the Point proportional tool with parameter t. Dragging this point along the curve will answer this question, displaying the values of t in the **Variables** toolbox.*

Q4. Does the curve have symmetry? Explain.

A4: The curve is symmetrical about y = x. If we replace x and y in the Cartesian equation for the curve, the equation does not change.

2. Verify your answer using the software.

 a. Choose **Draw** → Point and plot an arbitrary point *not on the axes or curve.* Select the point A and choose **Constrain** → Coordinate. Type (0,0) to plot a point O at the origin of the coordinate system.

 b. Click on the curve, choose **Construct** → Reflection.

 c. Click on the point O and then anywhere else to graph the line of reflection through the origin.

 d. Click on the line and choose **Constrain** → Slope. Type 1 in the open edit box.

*Note: Since the line of reflection must be a geometric object, the line y = x must be constructed geometrically, not as a function. The suggested method is one of the possible methods students can choose. They should observe that the reflected image of the curve coincides with the original curve. In order to be more convincing, students can change the color of the image and use **View** → Toggle Hidden to toggle between the curve and its reflected image.*

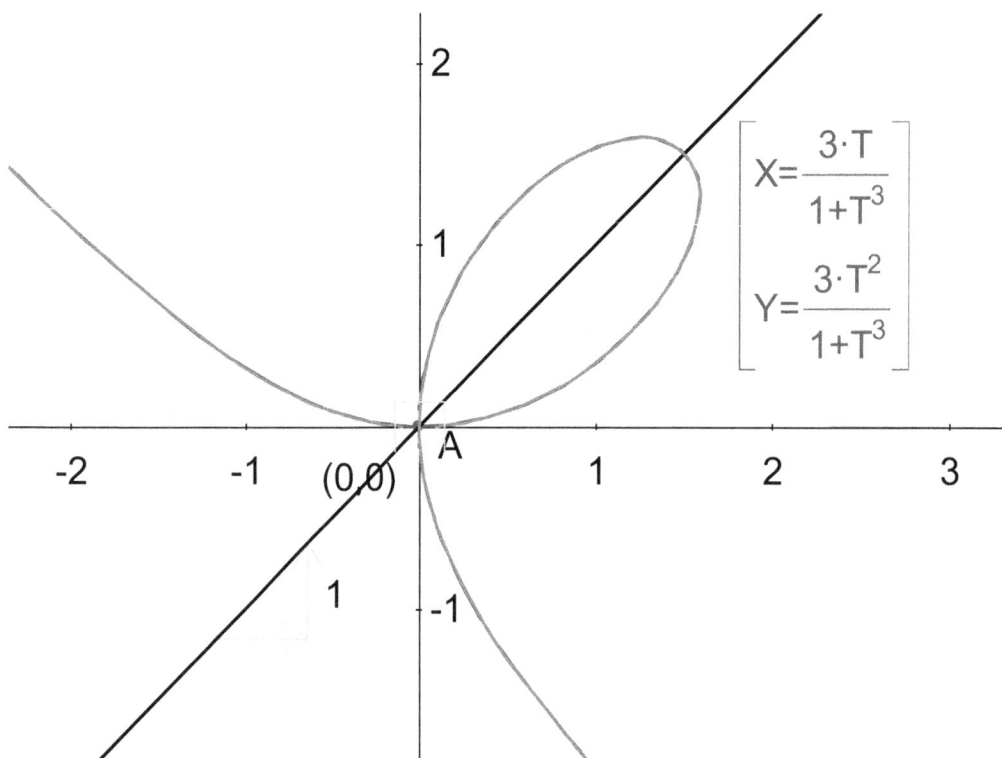

$$\left[\begin{array}{l} X=\dfrac{3\cdot T}{1+T^3} \\[4mm] Y=\dfrac{3\cdot T^2}{1+T^3} \end{array}\right]$$

EXPLORING ASYMPTOTIC BEHAVIOR

Q5. Is there a value of the parameter t where x and y are undefined? What happens to the values of x, y, and the ratio y/x when t approaches this value?

A5: *Consider the limits:* $\lim\limits_{t\to-1^+} x = \infty$ *and* $\lim\limits_{t\to-1^+} y = -\infty$; $\lim\limits_{t\to-1^-} x = -\infty$ *and* $\lim\limits_{t\to-1^-} y = \infty$.

Consider also the limit of the ratio: $\lim\limits_{t\to-1}\dfrac{y}{x} = \lim\limits_{t\to-1} t = -1$. *The infinite limits indicate that there is an asymptote, the limit of the ratio indicates that the asymptote has slope -1.*

3. Delete the line $y = x$ and the point you constructed at the origin.

4. Plot the function $y = -x - b$. Vary b and observe which line could be an asymptote for the curve.

   a. Choose **Draw** → Function. For the Type select Cartesian. Type the expression for the function in the prompt $Y =$.

   b. Drag the line and observe the values of b in the **Variables** panel.

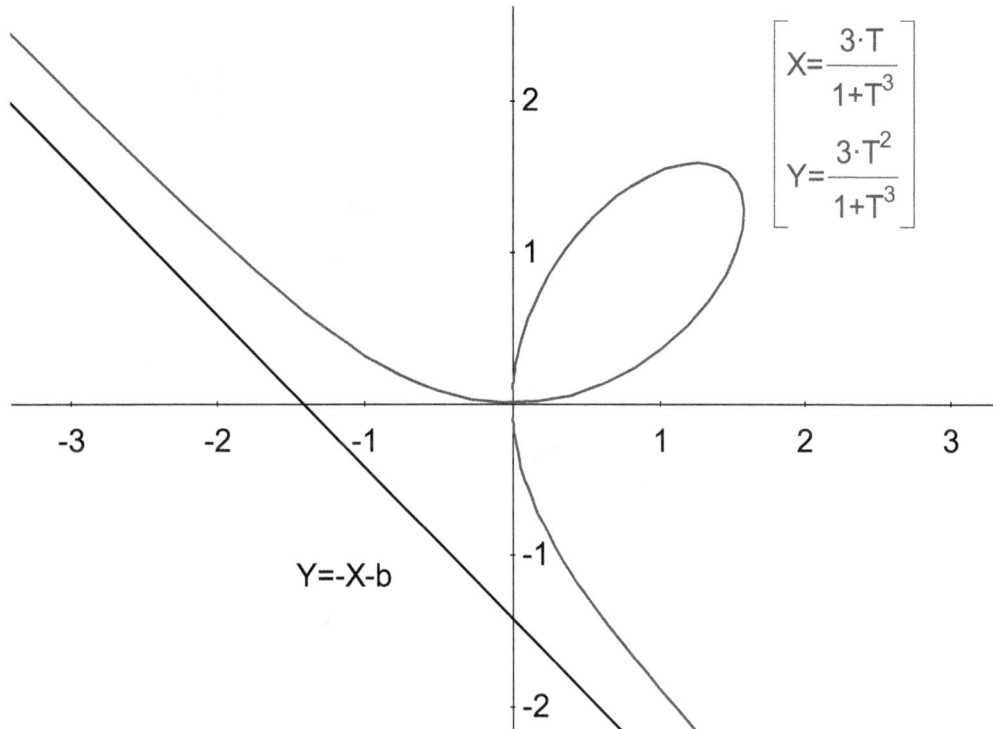

The curve is defined parametrically by

$$X = \dfrac{3 \cdot T}{1 + T^3}$$

$$Y = \dfrac{3 \cdot T^2}{1 + T^3}$$

and the line is labeled $Y = -X - b$.

Q6. Predict the value of b so that this line is an asymptote to the curve.

A6: *Students should guess that -b = -1 gives them reasonable guess for the value of b in the asymptote.*

5. Change the value of b *in your function* $y = -x - b$ according to your prediction.

   a. Double-click on the expression for the function and replace $-b$ with -1 .

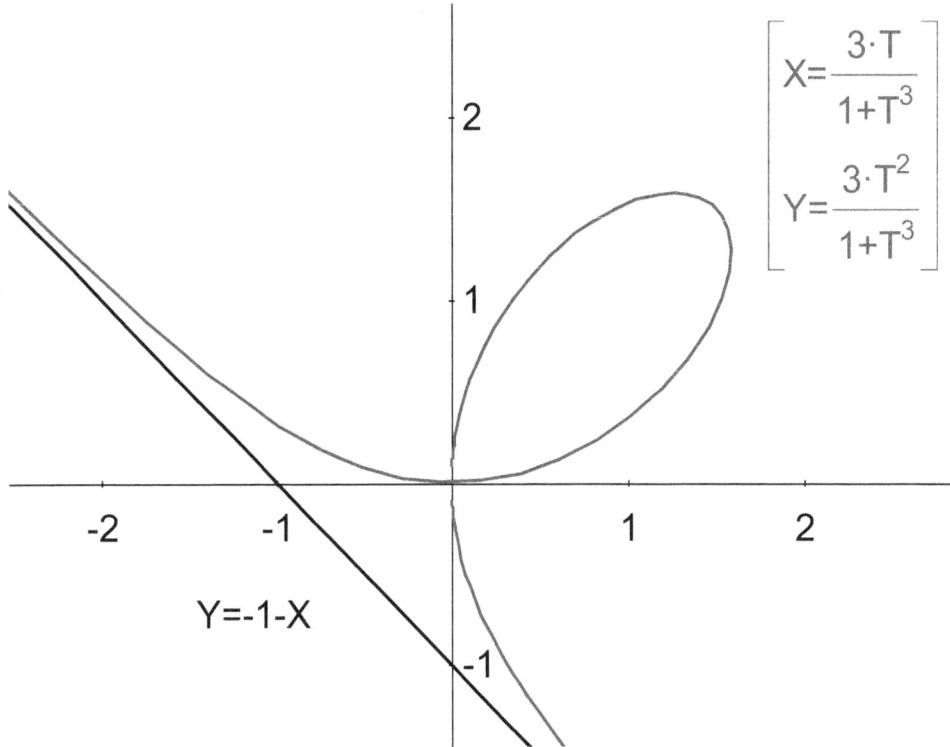

$$\begin{bmatrix} X = \dfrac{3 \cdot T}{1 + T^3} \\[4mm] Y = \dfrac{3 \cdot T^2}{1 + T^3} \end{bmatrix}$$

Y=-1-X

Q7. What happens to the distance between the points on the curve and the asymptote as you move away from the origin?

A7: The curve gets closer to the line.

Q8. How can you determine if this line is in fact an asymptote to the curve?

A8: If the distance between the curve and the line approaches zero when x and y approach infinity (or when t → -1), then the line y = -x – 1 is an asymptote to the curve.

6. Determine the distance between corresponding points on the curve and on the line. The corresponding points are chosen so that for an arbitrary value of the parameter *t*, the *x*-coordinates of the points are equal.

   a. Choose **Draw** → Point and plot a point on the curve. Select the point and the curve, and choose **Constrain** → Proportional along the curve. Type *t* in the open edit box.

b. Choose **Draw** → Point and plot a point on the line $y = -x - 1$. Select the point and the line, and choose **Constrain** → Proportional along the curve. Type 3*t/(1+t^3) in the open edit box.

c. Select both points, choose **Calculate** → Distance.

$$\left(\begin{array}{l} X = \dfrac{3 \cdot T}{1 + T^3} \\[4mm] Y = \dfrac{3 \cdot T^2}{1 + T^3} \end{array} \right)$$

Y=-1-X

$\dfrac{3 \cdot t}{1 + t^3}$

$\Rightarrow \left| \dfrac{(1+t)^2}{1 - t + t^2} \right|$

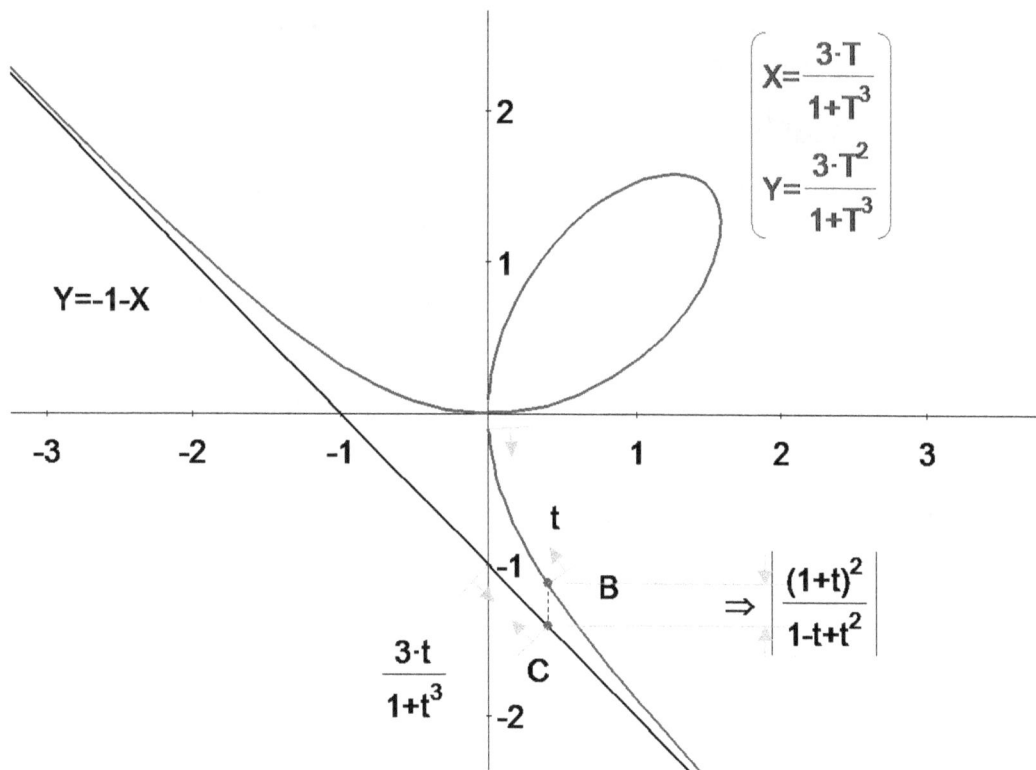

Q9. How does the distance between the point on the curve and the corresponding point on the line change as $x \to \pm\infty$?

A9: When $x \to \pm\infty$, $t \to -1$. At $t = -1$, the distance, BC = 0. Thus, $y = -x - 1$ is an asymptote.

Folium of Descartes Using Parametric Equations

Exploration 7.1: In geometry the Folium of Descartes is a plane curve described by the equation $x^3 + y^3 - 3axy = 0$. The name of the curve comes from the Latin *folium* which means leaf, and as you will see, the curve forms a loop like a leaf. Using parametric equations, you will investigate the folium of Descartes with $a = 1$.

EXPLORING SYMMETRY

Q1. Can you graph the Folium of Descartes using its Cartesian equation: $x^3 + y^3 - 3xy = 0$? Explain.

Q2. Find parametric equations for this curve. (Hint: use $y = xt$, where t is a parameter).

1. Draw the curve using parametric equations.

Q3. What values of parameter t correspond to the different parts of the curve?

Q4. Does the curve have symmetry? Explain.

2. Verify your answer using the software.

EXPLORING ASSYMPTOTIC BEHAVIOR

Q5. Is there a value of the parameter t where x and y are undefined? What happens to the values of x, y, and the ratio y/x when t approaches this value?

3. Delete the line $y = x$ and the point you constructed at the origin.

4. Plot the function $y = -x - b$. Vary b and observe which line could be an asymptote for the curve.

Q6. Predict the value of b so that this line is an asymptote to the curve.

5. Change the value of b in your function $y = -x - b$ according to your prediction.

Q7. What happens to the distance between the points on the curve and the asymptote as you move away from the origin?

Q8. How can you determine if this line is in fact an asymptote to the curve?

6. Determine the distance between corresponding points on the curve and on the line. The corresponding points are chosen so that for an arbitrary value of parameter *t*, the *x*-coordinates of the points are equal.

Q9. How does the distance between the point on the curve and the corresponding point on the line change as $x \rightarrow \pm\infty$?

7.2 Folium of Descartes in Polar Coordinates

<u>Exploration 7.2:</u> In this activity you will use polar coordinates to explore some properties of a plane curve known as the Folium (Leaf) of Descartes. We will consider the special case of the curve described by the equation $x^3 + y^3 - 3xy = 0$.

SUMMARY

<u>Mathematics Objectives:</u>

- Explore the tangent lines to the Folium of Descartes using polar coordinates.

- Determine the size of the leaf formed by the curve using polar coordinates.

<u>Vocabulary:</u>

- Pole

- Horizontal tangent

- Vertical tangent

<u>Pre-requisites:</u>

- Polar coordinates and polar equations

- Parametric derivative

- Derivative of trigonometric functions

- 1st derivative test

<u>Problem Notes:</u>

- The Folium of Descartes described by $x^3 + y^3 - 3xy = 0$ is a fascinating curve that is very difficult to graph in Cartesian plane. This curve has a double point in the origin, symmetry about the line $y = x$, and a slant asymptote $y = -1 - x$. The curve passes through the origin twice at $\theta = 0$ and $\theta = \pi/2$ creating a double point.

- Students first derive the polar equation using the given Cartesian equation of the curve and substituting $x = r\cos\theta$ and $y = r\sin\theta$ in order to graph the curve. They use the parametric derivative in order to explore the tangent lines to the curve at the pole.

- Students then use an optimization process to determine the maximum size of the leaf.

Technology skills:

- Draw: function, point

- Construct: tangent line

- Constrain: point proportional along a curve, coordinates of a point

- Calculate: distance, slope

Extension:

Explore tangent lines and find the size of the leaf of the Folium of Descartes for an arbitrary value of $a > 0$ using polar coordinates.

STEP-BY-STEP INSTRUCTIONS

EXPLORING TANGENT LINES AT THE POLE

Q1. Can you graph the Folium of Descartes using its Cartesian equation: $x^3 + y^3 - 3xy = 0$? Explain.

A1: *Since it requires solving a cubic equation and graphing three different solutions, this curve is very difficult to graph using a Cartesian equation. It is much easier to graph this curve using parametric equations or polar coordinates.*

Q2. Use $x = r\cos\theta$ and $y = r\sin\theta$ to find the polar equation for this curve.

A2: *Substitute the expressions for x and y into the equation,*

$$r^3 \cos^3\theta + r^3 \sin^3\theta - 3r^2 \cos\theta\sin\theta = 0 \Leftrightarrow r^2\left(r(\cos^3\theta + \sin^3\theta) - 3\cos\theta\sin\theta\right) = 0 , so$$

solving for r we get: $r = \dfrac{3\cos\theta\sin\theta}{\cos^3\theta + \sin^3\theta}$, *where* $0 \le \theta \le 2\pi$.

1. Draw the curve using the polar equation.

   a. Use **Toggle grid and axes** to show the axes without the grid.

   b. Choose **Draw** → Function. For the Type select Polar. In the r = prompt type the expressions for $r(t)$ you found. (Be sure to use T in place of θ.) Select 0 for Start and 6.28 for End and press OK.

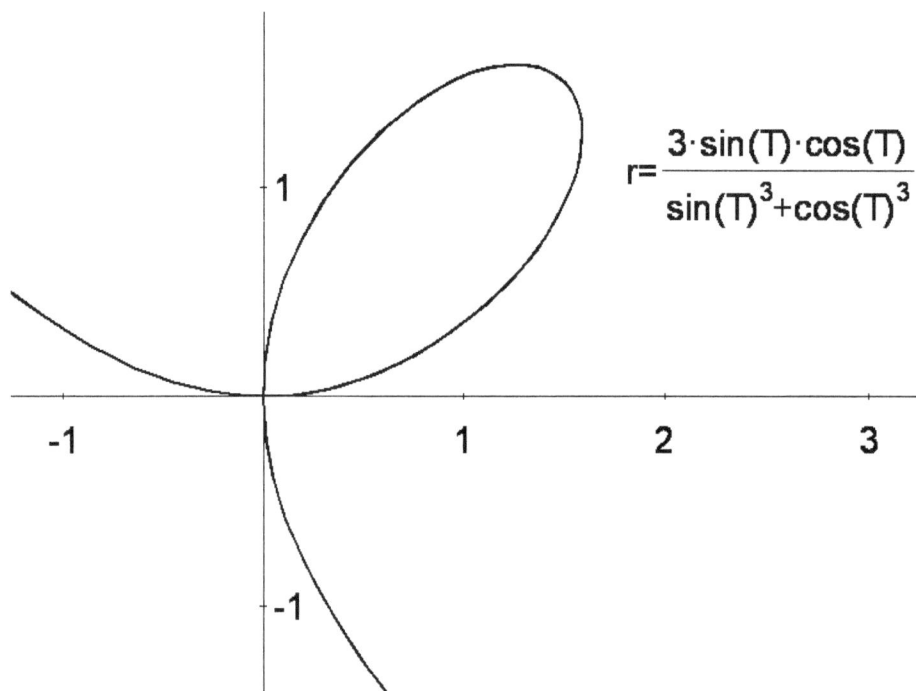

$$r = \frac{3 \cdot \sin(T) \cdot \cos(T)}{\sin(T)^3 + \cos(T)^3}$$

Q3. What are the tangent lines to the curve at the pole (0,0)?

A3: *The curve has two tangent lines, horizontal and vertical, so $\theta = 0$ and $\theta = \frac{\pi}{2}$.*

Q4. What must be true for the curve to have a horizontal tangent line? A vertical tangent line?

A4: *For a horizontal tangent line, $\frac{dy}{dx} = 0$. For a vertical tangent line $\frac{dy}{dx} \to \pm\infty$*

2. Explore the slope of the tangent line with the help of *Gx* and determine if the curve has vertical and horizontal tangent lines at the pole.

   a. Select the curve and choose **Construct** → Tangent. The point of tangency A will be displayed.

   b. Select point A and the curve, and choose **Constrain** → Point proportional along curve. Type *t* in the open edit box.

   c. Select the tangent line and choose **Calculate** → Slope.

d. Select the expression for the slope, and choose **Edit** → Copy As → String.

e. Choose **Draw** → Function. Select Parametric as a Type, in the X= prompt type *t*. In the Y= prompt paste expression for the slope using Ctrl – V. Choose Start = 0, End = 1.57, and press OK.

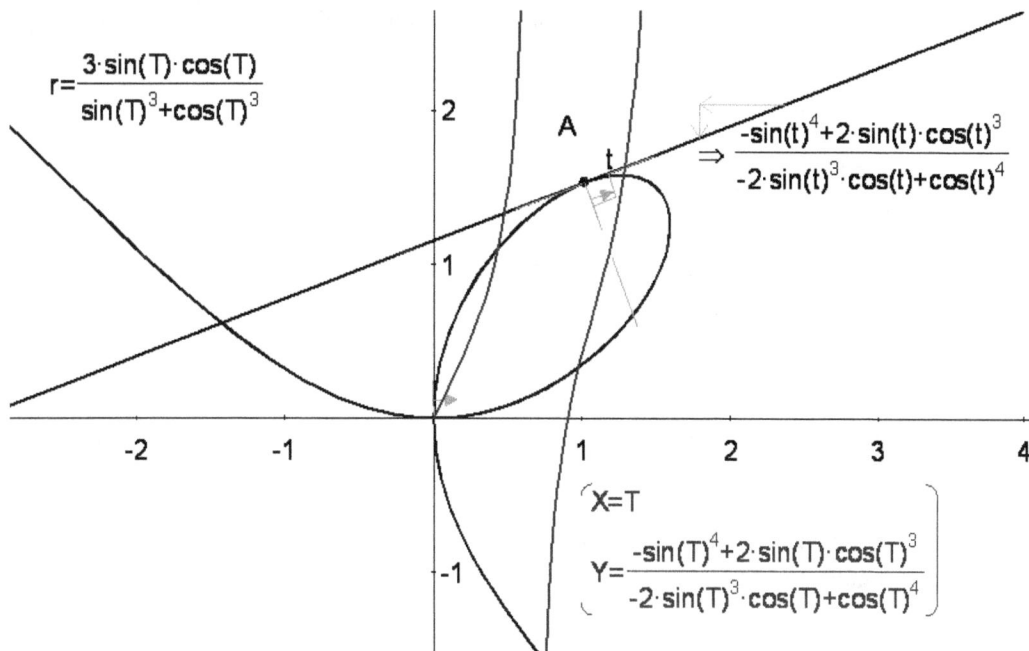

$$r=\frac{3\cdot\sin(T)\cdot\cos(T)}{\sin(T)^3+\cos(T)^3}$$

A

$$\Rightarrow \frac{-\sin(t)^4+2\cdot\sin(t)\cdot\cos(t)^3}{-2\cdot\sin(t)^3\cdot\cos(t)+\cos(t)^4}$$

$$X=T$$

$$Y=\frac{-\sin(T)^4+2\cdot\sin(T)\cdot\cos(T)^3}{-2\cdot\sin(T)^3\cdot\cos(T)+\cos(T)^4}$$

Q5. Explain your answers to questions 3 and 4 based on the results provided by the software.

A5: *Based on the graph, at the pole the slope is equal to zero when t = 0, and it is undefined when $t=\frac{\pi}{2}$, so the curve has horizontal and vertical tangents. Analytically, using the equation for the slope provided by the software we get* $\frac{dy}{dx}=\frac{\sin t(2\cos^3 t-\sin^3 t)}{\cos t(\cos^3 t-2\sin^3 t)}$.

Substitute points of interest: $\frac{dy}{dx}\Big|_{t=0}=0$ *and* $\frac{dy}{dx}\Big|_{t=\frac{\pi}{2}}=undefined$.

EXPLORING THE SIZE OF THE LEAF

Q6. For what value of t does the curve go through a point in the 1st quadrant that is the greatest distance from the origin?

A6: Due to symmetry, this should happen when $t = \dfrac{\pi}{4}$ and then

$$r = \frac{3\cos\left(\dfrac{\pi}{4}\right)\sin\left(\dfrac{\pi}{4}\right)}{\cos^3\left(\dfrac{\pi}{4}\right) + \sin^3\left(\dfrac{\pi}{4}\right)} = \frac{3\sqrt{2}}{2}$$

3.　　Calculate the distance from the pole to this point.

　　a.　Delete everything except the Descartes curve and its equation from the diagram.

　　b.　Choose **Draw** → Point and plot a point A on the curve.

　　c.　Select point A and the curve, and choose **Constrain** → Proportional along the curve. Type $\pi/4$ in the open edit box.

　　d.　Plot point B at the origin by choosing **Draw** → Point. Then constrain its coordinates to (0,0).

　　e.　Select the points A and B and choose **Calculate** → Distance.

$$r = \frac{3 \cdot \sin(T) \cdot \cos(T)}{\sin(T)^3 + \cos(T)^3}$$

$$\Rightarrow \frac{3 \cdot \sqrt{2}}{2}$$

$$\frac{\pi}{4}$$

A

B

(0,0)

-2   -1   1   2

2

1

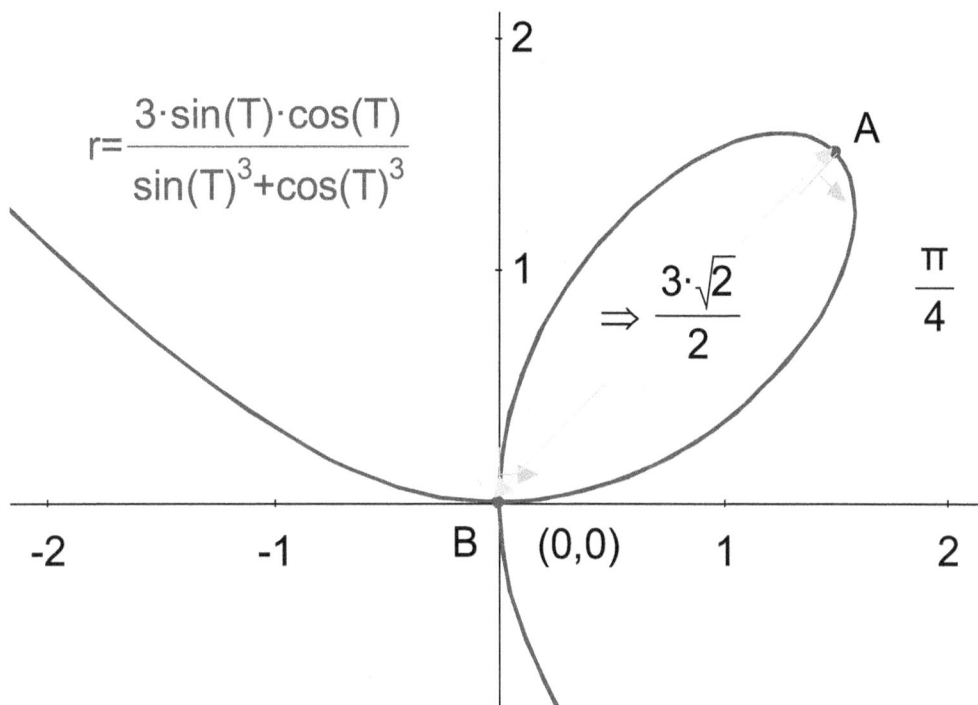

4. Use the software confirm that the maximum of $r(t)$ exists.

   a. Select the expression for the function $r(t)$, and choose **Edit** → Copy As → String.

   b. Choose **Draw** → Function. Select Parametric as a Type, in the X= prompt type t. In the Y= prompt paste expression for the slope using Ctrl – V. At the beginning of the expression delete $r =$.

   c. Choose Start = 0, End = 1.57, and press OK.

   d. Choose **Draw** → Point and plot an arbitrary point D. Select point D and choose **Constrain** → Coordinates. Type $\left(\frac{\pi}{4}, \frac{3 \cdot \sqrt{2}}{2} \right)$. The point will be plotted on the graph at the point that is maximum.

$$r = \frac{3 \cdot \sin(T) \cdot \cos(T)}{\sin(T)^3 + \cos(T)^3}$$

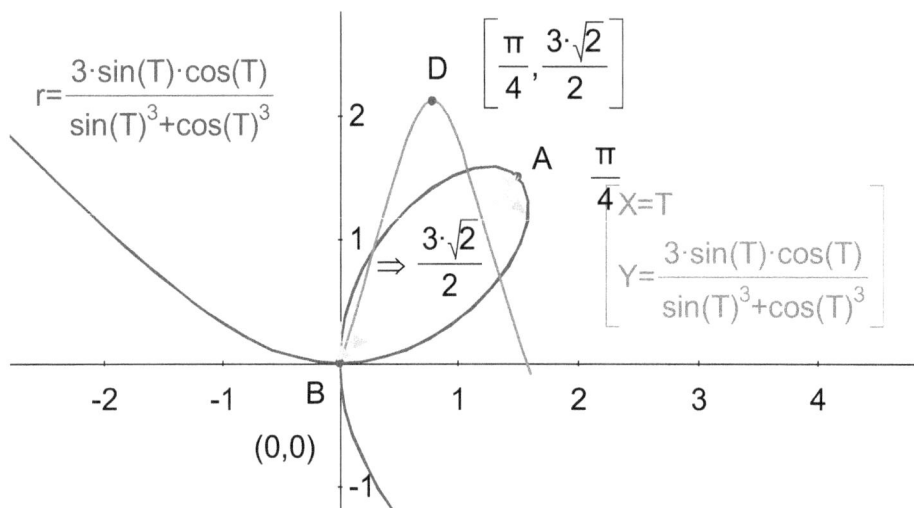

$$\left[\frac{\pi}{4}, \frac{3 \cdot \sqrt{2}}{2} \right]$$

A $\frac{\pi}{4}$

$$\frac{3 \cdot \sqrt{2}}{2}$$

$$X = T$$
$$\left[Y = \frac{3 \cdot \sin(T) \cdot \cos(T)}{\sin(T)^3 + \cos(T)^3} \right]$$

(0,0)

Q7. Confirm that the maximum of $r(t)$ exists using calculus.

A7.

$$\frac{dy}{dx} = 3 \frac{(\cos^2 t - \sin^2 t)(\sin^3 t + \cos^3 t) - (3\sin^2 t \cos t - 3\cos^2 t \sin t)}{(\sin^3 t + \cos^3 t)^2}$$

$$= 3 \frac{(\cos t - \sin t)\left((\cos t + \sin t)(\sin^3 t + \cos^3 t) + 3\sin t \cos t\right)}{(\sin^3 t + \cos^3 t)^2} = 0$$

Students can see that the second term in the numerator of derivative is always positive on the interval $\left[0, \frac{\pi}{2} \right]$, *and the denominator is always positive, thus,* $\cos(t) - \sin(t) = 0$ *at* $t = \frac{\pi}{4}$ *and changes sign from positive to negative.*

If using software to find the derivative, students could plot a tangent to a graph of the parametric function at a point with coordinate t, calculate the slope of the tangent line, and then substitute $t = \frac{\pi}{4}$ *to the expression for the slope to confirm that dy/dx=0:*

$$\left. \frac{dy}{dx} \right|_{\frac{\pi}{4}} = \left. \frac{3\left(-\sin^5 t - 2\sin^3 t \cos^2 t + 2\sin^2 t \cos^3 t + \cos^5 t\right)}{(\sin^3 t + \cos^3 t)^2} \right|_{\frac{\pi}{4}} = 0$$

Folium of Descartes in Polar Coordinates

Exploration 7.2: In this activity you will use polar coordinates to explore some properties of a plane curve known as the Folium (Leaf) of Descartes. We will consider the special case of the curve described by the equation $x^3 + y^3 - 3xy = 0$.

EXPLORING TANGENT LINES AT THE POLE

Q1. Can you graph the Folium of Descartes using its Cartesian equation: $x^3 + y^3 - 3xy = 0$? Explain.

Q2. Use $x = r \cos \theta$ and $y = r \sin \theta$ to find the polar equation for this curve.

1. Draw the curve using polar equation.

Q3. What are the tangent lines to the curve at the pole (0,0)?

Q4. What must be true for the slope of the curve to have a horizontal tangent line? A vertical tangent line?

2. Explore the slope of the tangent line with the help of software and determine if the curve has vertical and horizontal tangent lines at the pole.

Q5. Explain your answers to questions 3 and 4 based on the results provided by the software.

EXPLORING THE SIZE OF THE LEAF

Q6. For what value of t does the curve go through a point in the 1st quadrant that is the greatest distance from the origin?

3. Calculate the distance from the pole to this point.

4. Use the software to confirm that the maximum of $r(t)$ exists.

Q7. Confirm that the maximum of $r(t)$ exists using calculus.

Conclusions

The main feature of this book is that it does not focus on "how to do calculus" as do most secondary school curricula, but it focuses on the conceptual understanding of major calculus ideas. It is vitally important to make sure that students actually understand calculus and not just memorize algorithms. Going through the motions and mechanics of calculus does not ensure that students really know the concepts. The major complaint of many college calculus professors is that students who had calculus in high school, even at an AP level with a score of 5 on the AP exam, come to college not knowing calculus because they really do not understand it.

This book provides 29 interactive explorations that utilize the constraint-based dynamic software *Geometry Expressions* for teaching and learning many of the major ideas of calculus. The emphasis of the book is on two major concepts of calculus: the derivative and the integral. In these explorations we demonstrate how this software helps students to learn the theoretical ideas of calculus as well as to solve specific calculus problems.

Using *Geometry Expressions* provides learning opportunities for both teacher and students. The use of modern technology in teaching mathematics is an important part of current mathematics education, and is endorsed by the NCTM and the College Board. We believe that learning and appropriately using new educational computer software should be part of a teacher's ongoing professional growth.

Exposing students to dynamic visualization in learning calculus provides them with the opportunity to

a. Understand theoretical statements of calculus at a deeper level.

b. Develop investigation skills in the problem solving process.

c. Verify their own reasoning and steps of problem solving.

d. Develop a new view of mathematics as an experimental subject.

This software also allows students to analyze general symbolic expressions facilitating the idea of computer proof, *i.e.* justification of general mathematics ideas with the help of a computer.

We hope that every reader will have encountered something fruitful in these problems. Perhaps you've discovered new approaches to a familiar calculus problem or seen fresh connections between traditional content areas or learned entirely new calculus concepts. We also hope that you've been challenged to reflect on the role that using constraint-based computer software such as *Geometry Expressions* should play in teaching calculus and how this software facilitates students' ability to think critically and construct proofs.

Appendices

Appendix A: *Geometry Expressions* Keyboard Shortcuts

Command	Windows	Mac
File Menu		
New	Ctrl + N	⌘ + N
Open	Ctrl + O	⌘ + O
Close	Ctrl + W	⌘ + W
Save	Ctrl + S	⌘ + S
Print	Ctrl + P	⌘ + P
Edit Menu		
Undo	Ctrl + Z	⌘ + Z
Redo	Ctrl + Y	⌘ + Y
Select All	Ctrl + A	⌘ + A
Cut	Ctrl + X	⌘ + X
Copy	Ctrl + C	⌘ + C
Paste	Ctrl + V	⌘ + V
Delete	Del	⌫
View Menu		
Hide	Ctrl + H	⌘ + H
Zoom In	Ctrl + =	⌘ + =
Zoom Out	Ctrl + -	⌘ + -
Scale Geometry Up	Alt + =	Option + =
Scale Geometry Down	Alt + -	Option + -

Command	Windows	Mac
Help Menu		
Dynamic Help	Ctrl + F1	⌘ + F1
Contents	Ctrl + Alt + F1	⌘ + Option + F1
Index	Ctrl + Alt + F2	⌘ + Option + F2
Search	Ctrl + Alt + F3	⌘ + Option + F3

Brown, T. (2009) *Function Transformations.*

This unit is designed to familiarize students with the ideas of how various functions can be transformed, and the effect those transformations have on equations, graphs, and contextual situations. It specifically investigates three basic function types: those with parent functions y=x^2, y=1/x, and y=sin(x). These function families were chosen because they are simple enough for students to readily understand, and also sophisticated enough to clearly demonstrate the effects of dilations and translations. The skills and principles to be learned apply to virtually all function families, and will give a solid foundation for more advanced studies in functions.

Lyublinskaya, I., Ryzhik, V., & Funsch, D. (2009) *Developing Geometry Proofs with Geometry Expressions.*

We wrote this book to help teachers engage their students in learning geometry and encourage them in their development of logical thinking. This book offers 43 interesting geometry problems that resulted from a collaboration of Russian and American educators. They reflect the content of the standard high school curriculum. Altogether these problems demonstrate the most important methods of geometry. The unique features of these problems include the use of the symbolic geometry software, *Geometry Expressions*. Each problem includes student black line masters and detailed teacher notes with solutions, proofs, and steps for working with *Geometry Expressions*.

Lyublinskaya, I., Ryzhik, V. (2008) *Using Symbolic Geometry to Teach Secondary School Mathematics - Geometry Expressions activities for algebra 2 and pre-calculus.*

Our goal was to use *Geometry Expressions* to develop a set of problems for second year algebra and/or precalculus courses, addressing the topics of geometric transformations of functions and optimization. This volume contains eight interactive problems with different levels of difficulty. In each problem the main focus is on the development of the students' ability to recognize and make connections using multiple representations of the same object, such as geometric shape and function. By connecting geometric and algebraic representations the student develops a more thorough understanding of the problem.

Ottman, L. (2010). *The Farmer and the Mathematician: using Geometry Expressions and Google Earth to investigate crop circles.*

The typical student complaint about secondary mathematics is that it appears to have no practical application. The design and implementation of crop circles using center pivot irrigation provides us with an excellent example to see mathematics in action. By investigating these crop circles the lessons in this book utilize images from Google Earth, and a lighthearted scenario of a discussion between a farmer and a mathematician to introduce important mathematical concepts that are appropriate and adaptable for students in a range of courses from Pre-Algebra, Algebra, and Geometry, up through and including Calculus.

Shepard, I. (2008) *Exploring with Geometry Expressions in High School Mathematics*

Ian Shepard's book is based around a set of activities using the computer program *Geometry Expressions*. The author's passion to provide students with engaging learning opportunities is evident throughout. The activities embrace investigative approaches and seek to enhance student learning through discovery and the use of computer technology. *Geometry Expressions* has two features that add to the genre of dynamic interactive geometry software, constraints and symbolics. The activities and approach are designed to emphasize these additional features, constraints as an intuitive way to construct diagrams and symbolics as a link between algebra and geometry.

Todd, P. (2007) *101 Symbolic Geometry Examples Using Geometry Expressions.*

This book comprises 101 examples of the use of *Geometry Expressions* in a variety of settings. In some cases, we give a simple model with little explanation. In other cases, there is some exposition backed up by *Geometry Expressions* models. Some examples make use of an algebra system in addition to *Geometry Expressions*, while many use *Geometry Expressions* stand-alone. Together, we hope, they give an indication of what the system can do, and provide a starting point for the reader to pursue his own discoveries.

Todd, P. (2009) *101 Conic Sections Examples Using Geometry Expressions.*

The goal of this book is to show you what you can do with the conic section capabilities of *Geometry Expressions*. The goal is not to teach you conic sections, though you might learn some interesting and arcane facts about them on the way. Neither is the goal to teach you how to use *Geometry Expressions*, though you might pick up a few pointers en route. Many of the examples illustrated in this book are taken from the classic text *Conic Sections* by George Salmon.

Wiechmann, J. (2009) *Connecting Algebra and Geometry Through Technology - applying Geometry Expressions in the Algebra II and Pre-Calculus classrooms.*

Geometry Expressions provides a playground where students can discover their own mathematics. They will begin to see mathematics as something that is created, not just a set of facts made up long ago. Once students take ownership of their mathematics, they will be more apt to work productively and reflectively, with the skilled guidance of their teachers. The units presented in this book are a jumping-off point for using *Geometry Expressions* in the classroom. Use the units to gauge the potential of this powerful software, and as a guide to applying *Geometry Expressions* in your own classroom. We trust that you will enjoy using these units and the software.

Coming Soon:

Ottman, L. (2010). *The Tortoise and Achilles: Using Geometry Expressions to Investigate the Infinite.*